Activities Handbook for Teaching the Metric System

Gary G. Bitter

Arizona State University

Jerald L. Mikesell

Mesa Public Schools

Kathryn Maurdeff

Mesa Public Schools

Allyn and Bacon, Inc.
Boston, London, Sydney

LIBRARY OF CONGRESS CATALOGING IN PUBLICATION DATA

Bitter, Gary G
    Activities handbook for teaching the metric system.

    Bibliography: p.
    Includes index.
    1. Metric system—Study and teaching. I. Mikesell,
Jerald L., 1935-   joint author. II. Maurdeff,
Kathryn, joint author. III. Title.
QC93.B5   389'.152'071   75-17861

ISBN 0-205-04848-X

Third printing . . . May, 1976

Contents

76684

7 IDEAS   291

five-minute fillers; class activities; bulletin boards

8 THE METRIC SYSTEM IN THE UNITED STATES   309

history, metric historical dates

9 DESCRIPTION OF DERIVED UNITS   317

speed; velocity; acceleration; force; work; power; energy; pressure; density; specific gravity; mole; degrees kelvin; heat; light

10 CAREERS AND METRIC   321

Preface

This book is directed to the student, teacher, or layman who wishes to become familiar with the metric system and its uses in everyday life. The reader/participant will attain an understanding of and a facility in the use of the metric system. The material in this book can be used as a resource for workshops, inservice courses, individual study, or as a reference for teachers. Numerous activities are available for the exploration and understanding of the metric system, as well as drill and practice in each of the following categories: linear measurement (distance), volume and capacity, area and perimeter, temperature, and mass (weight).

MEASUREMENT AND THE METRIC SYSTEM

The teaching of measurement is discussed along with the implications of Piaget's research relating to measurement. Indirect and direct measurement, approximate measurement, and standard and nonstandard unit topics are also included. Suggested metric curricula for primary, intermediate and junior high, and secondary grades are outlined.

INSTRUCTIONAL STRATEGIES

The background materials and the suggested activities can be set up for individual, small-group, or large-group instruction. Suggestions for activities appropriate for adult education, PTA groups, inservice training, and primary and intermediate grade instruction are outlined. The activities are designed to give ideas for class presentations, interest centers, or individual activities. These activities can be used for daily or weekly curricula, or can be extended over the entire academic year at any particular level. Keep in mind that if the person is not familiar with the metric system, introductory activities are essential. The most commonly used metric units are included in the activities.

MATERIALS NEEDED

Readily available materials can be used for most of the activities. However, it would be helpful to have some commercial materials for the participants as they become more proficient with the metric system. The activities have been class-tested. Hopefully, the activities selected by the teacher will be those that are appropriate for his or her particular grade or group.

SI METRICS

In an effort to have a truly international system of measurement, *The International System of Units* (abbreviated SI) was established. The parameters of the SI system are discussed at length.

ACTIVITIES

The activities are considered the core of this book. Each activity includes:

Teaching Objectives—Each activity has behaviorally stated objectives. The teacher is told what the student should be able to do as a result of successfully performing each activity.

Problem—The activity is stated in terms of a problem. These are excellent headings or questions for activity cards.

Appropriate Group Size—Indications are given as to the ideal number of participants who could be involved in the activity.

Materials—Materials needed for each activity are delineated.

Discussion—Key questions are asked to stimulate thinking and mental involvement in the activity.

Prerequisites—Skills necessary for successful participation in each activity are spelled out.

Directions—Step-by-step instructions make individualization a reality.

Teaching Notes—Suggestions for background materials and ideas are valuable aids to the teacher who may not be especially familiar with the metric system.

Extensions—Ideas which can turn all of the activities into open doors for exploration are given. Some are appropriate for all ability levels.

TEACHER HELPS

Five-minute fillers, whole-class activities, and bulletin board ideas are included as a valuable resource for establishing a learning climate for the metric system.

HISTORY

The turbulent history of the metric system from Thomas Jefferson's involvement to the present is chronicled in easy-to-read form.

DERIVED UNITS

Many derived units are discussed to illustrate the application of the metric system to the world of science.

OTHER INCLUSIONS

Metric mastery and diagnostic tests, powers-of-ten review, career implications, and conversion tables are among the additional materials given. An extensive bibliography—including articles, books, and films—and a list of suppliers will hopefully make it easy to find additional metric sources to be used for further exploration and research.

It is the hope of the authors that the readers will find all they need to know about the metric system in this book. Good luck and we hope you find it

"A Pleasure to Measure Metrically"

1

Introduction

The task of retooling for any change as dramatic and comprehensive as "going metric" is an enormous one. Every American is or soon will be facing the task of learning to THINK METRIC. Living and functioning in a metric America will cause some discomfort at first, but as almost anyone who has lived for a time in a nation which uses the metric system of measuring can tell you, it is possible to adapt quickly and quite easily.

It is evident by reading the history of the metric system that the time has come for the United States to make the change. Although many companies and industries are already changing, most of the "average citizenry" has not been greatly affected. It takes a tremendous amount of energy, creativity, courage, commitment, and willingness to take risks on the part of many people to accomplish any task as big as adopting the metric system.

As with any major movement there is a certain amount of resistance to change, but once people begin to "think metric" this resistance will fade and the merits of the system will have us wondering why we did not make the change many years ago.

Many people have heard vague references to parts of the metric system and may have a fair mental picture of a meter and a kilogram, but most people's experiences with the metric system have been limited to doing a few problems requiring converting to the customary United States system and back. This process is laborious and, for most people, very unnecessary. It is important to have general ideas of the relationships that do exist, but learning how long a centimeter is and how warm 25°C is becomes the important task of anyone who lives in a country that uses the metric system.

Through practical work and physical as well as mental involvement, people become conversant with the metric units of measure, understanding their relationships and uses. The activities and exercises presented in this book are designed to get the user both physically and mentally involved. Emphasis is placed on understanding the relationships that exist and on obtaining a valid mental picture. Once these concepts are firmly established, *then* drill and practice activities are provided in abundance to develop mastery.

for the individual

For the reader who wants to become knowledgeable and competent in the metric system, this book serves as a comprehensive resource which will meet the needs of nearly everyone.

for the classroom teacher

There are many metric ideas in this book from which to choose. The activities are all organized into one chapter. This makes it possible for the teacher to select activities which can complement existing materials and curricula.

The prerequisite skills needed for each activity are listed so that the teacher can guide students with all levels of ability into meaningful activities in each measurement area.

The learning activities can meet the needs of individuals, small groups, and whole classes. Most of the activities also would provide valuable experiences for interest centers.

The activities are structured to some degree, but it is possible to vary from the described procedure and to set up alternative ways to solve the problem. The activities are arranged in such a way as to require the student to solve the experiment or activity in an organized way, including the collection of data when appropriate. Answers are purposely omitted and hopefully the student will experiment through trial and error to reach a meaningful conclusion or result. Since the activities include *exploration, discovery,* and *drill and practice* activities, the teacher should make an appropriate choice of each to establish a meaningful sequence, with the drill and practice as the final activities. Be sure to allow adequate time for the students to complete the activities. Often activities are too hurriedly done and the purpose of the activity is not achieved. Encourage experimentation and then discuss answers which do not make sense. Hopefully, through careful teacher guidance and discussion, errors or misconceptions can be eliminated.

workshops, inservice education, and courses

This book is also designed to meet the need for metric activities and background for use in workshops, inservice education, and classroom instruction. A pretest or diagnostic test (Appendix B) can be administered to the group to determine general metric competency. From these results appropriate activities can be selected.

The activities can be set up conveniently for workshops by selecting the desired activities, placing each activity on a card, and providing all the necessary materials. Set up stations covering *linear measurement, area, perimeter, volume, capacity, mass, temperature,* and *games.* (See Figure 1.) The participants could then rotate through the stations in an organized manner. No more than six participants should be in each group. Time limits should be set for each rotation.

Inservice courses can be handled in a similar way except that the rotation would be on a class or period basis.

general features

In addition to the activities section, there are diagnostic and mastery tests that will aid the student in determining areas that need special atten-

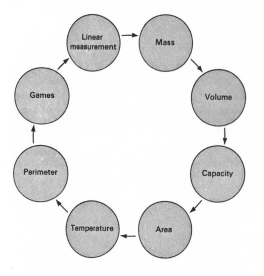

FIGURE 1. Workshop station rotation scheme.

tion. A brief historical development is presented as well as information regarding the International System of Units (SI). The inclusion of conversion tables, an extensive bibliography, and many other features make this book a valuable resource in all areas relating to metric measure.

2

A Metric America

America has "gone metric" in many aspects of daily living, although much of the population is unaware of it. Suddenly we are finding that the metric changeover is not as difficult as first thought. Once people have become acquainted with the metric system, they find the metric units similar to the customary units they replace. For example, the customer can buy a meter of fabric instead of a yard, a half kilogram of butter instead of a pound, a liter of milk instead of a quart; and the body temperature can be measured at 37°C instead of 98.6°F. As we become more familiar with metric names and build personal frames of reference, the metric system easily becomes a common part of our lives.

America is metric in many areas. Most measurements related to medicine are measured in metric units. The symbols cm^3, ml, and g are commonly used by the pharmaceutical industry. Cameras have used 8-, 16-, and 35-mm film for years. International sports are measured in metric distances.

Auto mechanics daily work on cars which have metric sizes. Optical lenses are determined by metric linear measure. The electrical industry uses metric units to measure electrical usage. Ordering wine in restaurants is a metric experience. And many items that you purchase are listed in metric units as well as in customary units.

Many major companies are changing to the metric system over the next few years. Some are as follows:[1]

1. International Business Machines
2. General Motors
3. Ford Motor Company
4. Caterpillar Tractor
5. International Harvester
6. Honeywell
7. NASA (spaceflight data)
8. 3M
9. John Deere
10. States with metric activities including road signs and education programs.

What are the advantages of the metric system?

1. It is easy to use.
2. It is a universal system.
3. It is based on 10, similar to our decimal number system as well as our monetary system.
4. The notation is systematic and easily interchangeable.
5. It is logical.
6. It is necessary for world trade.
7. It is easier to teach and learn.

What are the disadvantages of the metric system?

1. Not everything is metric.
2. People are familiar with the customary system.
3. Why change when people are happy now?
4. The changeover to metric will be expensive.

[1] This list is growing daily.

So, with more industries and educational programs going metric it is apparent that the metric system is here to stay and its advantages are many.

GO METRIC.   THINK METRIC.

reference

A Metric America: A Decision Whose Time Has Come. Daniel V. DeSimone. Washington: Government Printing Office, 1971.

3

Measurement

Measurement is one area of mathematics that has life-long implications since there are very few human activities that do not require some knowledge of measuring.

The speed limit is 80 km/hr.
The plane arrives at 3:00 p.m.
I bought 2 kg of potatoes.
The recipe called for 500 ml of milk.
She needed 2 m of material for the dress.

While measurement cannot be the only method used to teach computational skills, it does give students reasons for adding, subtracting, multiplying, dividing, etc. It can be said that measurement is one practical application of arithmetic to everyday life.

Much of our daily lives involves measuring. Therefore, it is relatively easy for the creative teacher to provide a variety of first-hand experiences in measuring.

Measurement enables us to compare or to describe items in our environment. The process of measuring involves the selection of a proper unit and the comparison of an item with that unit. The number of units that compose an item is its measure. The result will include the *number* of times the unit is used and the *name* of the unit (e.g., 4 meters).

The unit of measure selected is arbitrary but must be of the same nature as the item being measured (for example, a line segment is the only unit of measure for measuring distance). Measuring is a process of comparing two or more items and the *resulting* measurement *describes* the item.

direct and indirect measures

Measurement can be either direct or indirect. Some items can be measured directly by selecting a standard unit and comparing the unit directly to the item (e.g., length of a table, height of a child, etc.).

Indirect measurement is often used when it is not possible to place the measuring instrument directly on the item being measured, (e.g., height of a tree, area of a forest, temperature, distance to the sun, etc.) or for finding area or volume. If direct measurement is not feasible or possible, the objects are measured by comparing them to something (such as a shadow) that can be measured. Indirect measurement is often used because it is convenient.

measurement is approximate

The measurement of an object will always be approximate, no matter what unit of measure is selected. There will always be a shortage or excess when the unit of measure is compared to the object being measured. By using smaller and smaller units the precision of the measurement can be increased. But no matter what unit is chosen, it would always be possible to determine a smaller unit of measure.

The unit of measure chosen is determined by the degree of precision desired. For example, to measure distance between cities, the unit usually chosen is the kilometer; to measure the length of a room, the meter or centimeter is usually used; while to measure machine tolerances, the millimeter is an appropriate unit.

It is assumed that a measurement is as precise as the smallest unit reported. If an object is said to have a length of 10.5 m, the last digit is tenths and therefore the length has been reported to the nearest tenth of a meter. The actual measurement is anywhere from 10.45 to 10.55 m.

Convention has allowed the use of the equal sign between units of measure although measurements are approximate and really not equivalent. We also tend to ignore the fact that measurement is approximate in utilizing it in our everyday life. For example, we use digits connected with measurement as if they were exact (e.g., Sonia is 139 cm tall, Steven has a mass of 35 kg) and use nonstandard units of measure as if they were standard (e.g., a handful of flour, 4 blocks away, etc.). Therefore, all measurements are approximate and as precise as we define them.

nonstandard and standard units

The system of weights and measures has developed within our society as dictated by our needs.

Since early man apparently did not need a precise system of measurement, the first measurements were probably general comparisons of nonstandard units.

For example:

> The animal is larger than the child.
> It is heavier than this rock.
> I need more than that.

As societies became more complex, numeration systems evolved, making it possible to develop systems of weights and measures based on standard units that were suitable for a more complicated way of life. In developing these systems of measurement, early man utilized what was readily available as instruments of measure, namely his body and his environment. The systems that developed were fairly effective for a particular society's use.

However, as man began to exchange goods with his neighbor, a more precise system of measurement was needed. Units of measure were standardized first by small groups of people and then by the government of a country. Soon, a standard unit of measure was chosen and this measure was then used to measure other objects.

Therefore, a standard unit is one that is established by law, is precisely defined, and is used by members of the community in their everyday life. In addition to the standard unit, multiples and subdivisions are determined to measure smaller and larger quantities.

In the United States, the Constitution gives to Congress the right to establish standard units. In 1866 Congress legalized the metric system. By 1893 the established national standards of length and mass became the meter and kilogram. The yard is defined in terms of the meter and the avoirdupois pound is defined in terms of the kilogram.

the metric system

The metric system is a highly organized system of measurement in which the units of length, area, volume, mass, and capacity are related. It has a close relationship to our decimal system of numeration since the units are expressed in powers of 10. Multiples and subdivisions of units are formed by multiplying or dividing by 10. This makes it relatively easy to change from one unit to another by shifting the decimal point.

4

The Teaching of the Metric System

go metric?

The changeover to the metric system of measurement in the United States seems inevitable, making the discussion of the pros and cons of the advisability of adoption irrelevant. The real questions become: (1) How do we go about accepting the metric system of measuring? and, (2) How do we prepare our students to live in a metric world?

Teachers will need to provide metric explanations and learning experiences for adults as well as children since the metric system will definitely affect their lives as well as the lives of their children. Presently, many adults are apprehensive about working with the metric system because it is an unknown quantity and their limited exposure to the system has been one of memorizing conversion tables and doing long and tedious computations based on these tables. Since many teachers have the same feelings about the

13

metric system it will be very important for them to learn, understand, and appreciate the metric system before trying to teach it. Using the metric system is not a "new" way of measuring. The amounts being measured do not change, just the terminology used to describe the amount will be new.

The concepts that (1) measurement is approximate, and (2) units of measure must be of the same nature as what is being measured, are often overlooked in the teaching of measurement. Hopefully as we begin to introduce the school population to the metric system these concepts will not be overlooked.

The teaching of the metric system brings about questions as to the pedagogical arrangement of topics in the mathematics curriculum. Since the metric system is based on powers of 10, there will be more emphasis on the teaching of decimals to children at an earlier age. Also more attention will be given to the teaching of place value and extending the youngsters' knowledge of working with the powers of 10. This, in turn, will bring about a decrease in instruction of fractions. Fractions more than likely will be included in the upper grades rather than their current placement in the curriculum.

Ideally, metric activities could be taught during the entire academic year as part of the regular mathematics curriculum. Or, if the teacher prefers, a unit on measurement could be periodically included. In either case, a hands-on approach is suggested, with students involved in a variety of educational exploratory and discovery experiences to help them learn and understand the metric system and measurement.

The teaching of metrics should preferably be taught as *another* measuring system. It is important that children be taught NO conversions, as most conversions from customary to metric and vice versa are difficult. During the change over to metrics in the United States, it would be wise to give students a framework of equivalent measures in the customary system. For example, knowing that a meter is slightly larger than a yard, a liter is slightly more than a quart, etc., will be helpful rather than knowing the precise difference.

Piaget and measurement

The teacher will need to carefully select appropriate measuring experiences to insure that the child is capable of performing and understanding the concepts.

Jean Piaget believes that measurement is a synthesis of the operations of subdivision into parts and of substitution of a part upon others. The ability to measure develops later than the number concept because it is more difficult to divide a continuous whole, such as an object being measured, into interchangeable subunits than it is to count a set of objects that are separate and discrete from each other, such as beads or blocks.[1]

Conservation, or the invariance of distance and length, is an essential prerequisite skill for measuring. How can a child measure if he thinks the ruler changes in length as it is moved? Measuring involves the ability to construct in the mind an independent reference system in space in which objects can be moved to new positions, but the length of the object in each position in space does not change. According to Piaget, children generally are able to conserve length sometime between six and eight years of age.

To the child who has not achieved conservation of the length of objects, an object becomes longer as it is moved away from him. It is difficult to think that measurement can be meaningful if the measuring instrument is thought to change length as it moves along an object to be measured. Once conservation is attained, Piaget concludes that the necessary intellectual concepts for measurement are present and the child can perform systematic measurement with understanding.

Piaget's work indicates that if systematic measurement is to be "taught" it should not be presented before the latter part of the third grade. This does not preclude a great deal of activity occurring prior to this time, where children make comparisons, learn the language of measurement (e.g., shorter than, longer than, further than, etc.), and arrange objects in a sequential order.

Piaget believes that the necessary concepts develop from within rather than from without for operational understanding. You cannot tell children how to measure; they should be provided with appropriate materials and be allowed to experiment and to try to solve measurement problems for themselves. The necessary concepts will develop when the child is old enough and when he or she is allowed to experiment with objects used in measurement.

The ability to measure area is again related to the child's ability to conserve. There must be the understanding that area does not necessarily change with shape.

> As a result of his experiments, Piaget concludes that the concept of conservation and the ability to measure occur at the same time for measuring one dimension (length) or two dimensions (area). Conservation of area is the outcome of being able to compare two areas that are of different shape on the

[1] Piaget, Jean, "How Children Form Mathematical Concepts," *Scientific American,* Nov., 1953, p. 78.

basis of counting (additive subdivision) the number of basic subdivisions in each, using a basic unit of measure such as a triangle or square.[2]

Attempts are often made to teach linear measurement before the child is at a readiness level to really understand it, but area measurement is deferred several years past the age when children can understand it. As soon as a child has achieved conservation (approximately age nine) he or she can begin measurement using a unit square and counting the number of times it is contained in the figure being measured. Using the rule for obtaining area by multiplying length by width is not appropriate until the age of eleven or twelve. This means that extended use of this method should be deferred until the fifth or sixth grade.

A study of volume measurement requires understanding the idea of conservation of occupied or displaced volume. A problem that would illustrate this level of conservation would be to have two cylinders that have identical shape and size but made of two different materials, one of which is heavy and the other one light. If the student can understand and explain why both displace the same amount of water, he or she is ready to understand the measurement of volume.

Children at about eight years of age can and should explore problems of volume using blocks or bricks, but the study of volume using the measurement formula is appropriate for the student aged eleven, twelve, or older.[3]

The measurement of mass or weight has developmental stages closely comparable with those of linear measure. There are many activities that are appropriate for children four years old and up, but these activities should involve comparing, classifying, ordering, and equalizing. A child should have attained conservation if measurement using standardized mass pieces is to be meaningful.

Anyone providing measurement-learning experiences needs to keep two important ideas in mind: (1) measurement is by nature a hands-on activity in which children need to get personally involved, (2) many concrete, exploratory activities should precede measurement using standardized tools or measurement algorithms. Following the progression from the con-

[2] Copeland, Richard W., *How Children Learn Mathematics,* London: The MacMillan Co., 1970, p. 236.
[3] Copeland, Richard W., *How Children Learn Mathematics,* London: The MacMillan Co., 1970, Chapters 10 and 12.

crete to the representational to the abstract is a good rule to follow when planning all measurement activities.

metric curriculum ideas

Metric-measurement activities can be an integral part of the mathematics curriculum at all grade levels. Exploring and discovering metric concepts by doing activities can involve students at all levels.

primary grades

The emphasis in the primary grades should be on *exploring* metric measures. Measurement activities will introduce the meaning of measurement, involve the students in simple problem-solving situations without utilizing standard units, help the students discover a need for standardized units, and introduce the common units of measure and the tools used to measure. First experiences help to develop measurement readiness. The students need to develop an intuitive understanding of some of the characteristics of measurement. They should be encouraged to invent processes for solving measurement problems. The process can be as important as the actual answer.

Listed below are *suggested* topics for the primary grades. They would not necessarily be covered in the order given.

Exploring metric measures

Matching, sorting, and comparing
Defining measurement
Nonstandard units
Ways measurements are used
Developing a standard measure
Estimating measurements
Relating metric system to numeration and monetary system
Measuring time and temperature
Measuring length—meter and centimeter
Measuring mass (weight)—kilogram
Measuring capacity—liter

Measuring perimeter
Measuring area—square unit

intermediate grades

The emphasis in the intermediate grades should be on *discovering* metric measures. Measurement activities can help students understand the importance of measurement in daily activities. The history of measurement could be discussed, measurement concepts refined, vocabulary extended, and students taught to use measuring instruments with increasing skill. In addition, new units could be introduced and many opportunities provided for students to work with the general relationships in the metric system.

Listed below are some *suggested* topics that could be covered in the intermediate grades. These topics would not necessarily be covered in the order given.

Discovering metric measures

The metric system
History of measurement
Approximate nature of measurement
Estimating volume and area
Decimal introduction of the measurement system
Decimal calculations and notations
Metric relationships between units
Changing units within the metric system
Linear measurement—millimeter, kilometer
Capacity measurement—milliliter
Mass (weight) measurement—gram
Area measurement—square meter, square centimeter
Volume measurement—cubic meter, cubic centimeter
Temperature measurement—Celsius

junior high and secondary grades

The emphasis in the junior and senior high school grades should be on *investigating* metric measures. Students need to be made aware of how measurements affect us both socially and economically. Vocabulary needs to be extended and refined. Students should have the opportunity to become familiar with accuracy, precision, and direct and indirect measurement.

Listed below are some *suggested* topics that could be covered in the junior and senior high school. These topics need not be covered in the order given.

Investigating metric measures

The metric system—what and why
The International System of Units
Direct and indirect measuring
Scientific notation
Significant digits
Accuracy and precision in measuring
Metric relationships between units
Decimal calculations
Linear measurement—(remaining units)
Volume measurement—(remaining units)
Mass (weight) measurement—(remaining units)
Temperature measurement
Capacity measurement—(remaining units)
Area measurement—(remaining units)
Measuring to scale

general references

Richard W. Copeland. *How Children Learn Mathematics.* London: The MacMillan Co., 1970.

John L. Philips. *The Origins of Intellect—Piaget's Theory.* San Francisco: W.H. Freeman and Company, 1969.

Jean Piaget. *The Origins of Intelligence in Children.* New York: W.W. Norton & Company, Inc., 1963.

5

The Do's and Don'ts of the Metric System (SI)

In order to have a truly international system of measure, *The International System of Units* (abbreviated SI in all languages) was established. SI is an agreed upon international system. Briefly, SI has established the rules, symbols, and style of the metric system to be used throughout the world. Some of the common rules of SI are as follows:

1. Avoid capitalization of unit names (except Celsius) unless they start a sentence.

   Example:
   meter, not Meter
   kilogram, not Kilogram

> *Note:* Unit names are not capitalized even though
> some of their symbols are, with the exception
> of degree Celsius.

2. Pluralization of symbols is not to be used.
    Example:
       3 mm, not 3 mms
       6 g, not 6 gs

3. Never use a prefix without a unit either in writing or speech.
    Example:
       kilometer or kilogram, not kilo
       millimeter or milligram, not milli

4. Use a zero before the decimal point when the numerical
    unit is a partial unit.
    Example:
       0.401 mm, not .401 mm
       0.5 g, not .5 g

5. Do not use periods with symbols except at the end of a
    sentence.
    Example:
       m, not m.
       cm, not cm.

6. When dividing, the use of an oblique stroke (/) is preferred,
    to separate the numerator and denominator.
    Example:
       meter per second squared—m/s^2
       kilogram per cubic meter—kg/m^3
       kilometers per hour—km/h

7. Prefixes in denominators are to be avoided, except with
    the kilometer. (Express denominators in terms of base
    unit, not multiples of it.)
    Example:
       MN/m^2, not N/mm^2

8. Commas are not to be used as place markers when writing
    large numbers. Instead use a space. The reason for this is
    that many countries using the metric system use the comma
    as we use the decimal point.
    Example:
       367 245.261 3, not 367,245.2613

9. Always leave a space between digits and symbols.

Example:
   67 m, not 67m
   0.123 cm³, not 0.123cm³

10. Prefixes that are powers of 10^3 (micro, milli, kilo) are preferred. Others should be avoided where convenient.
11. Avoid mixing multiples of units.
  Example:
   15.75 m, not 15 m 750 mm

what are the measurement units?

The International System of Units (SI) has defined seven base units. All of the base units, with the exception of the kilogram, are based upon a natural phenomenon which can be duplicated under laboratory conditions. For example the meter is 1 650 763.73 wave lengths in vacuum of the orange-red line of the spectrum of the krypton-86 atom. The use of these phenomena eliminates the need for international models as reference. But the kilogram is not a natural phenomenon. It is represented by a platinum iridium alloy cylinder kept in Paris at the International Bureau of Weights and Standards. The United States has a duplicate stored at the National Bureau of Standards.

TABLE 5-1. The International System of Units (SI).

Quantity	Base Unit	Symbol
Base Units		
length	meter	m
mass	kilogram	kg
thermodynamic temperature	kelvin	K
time	second	s
electric current	ämpere	A
amount of substance	mole	mol
luminous intensity	candela	cd
Supplementary Units		
plane angle	radian	rad
solid angle	steradian	sr

The other SI units are derived from the base units. The SI derived units are found by simple mathematical multiplication and division of the SI units without the introduction of any numerical factors.

Example of derived unit:

Area: square meter, m^2
Volume: cubic meter, m^3

Another action of SI was the establishment of agreed upon prefixes and symbols. They are given in Table 5-2.

The advantages of SI are as follows:

1. SI is a unique measurement system. Each quantity has only one unit associated with it (e.g., distance—meter).
2. SI is a simplified system. It is based upon the decimal system (like the U.S. monetary system and numeration system). Therefore, most calculations involve moving a decimal point either to the right or left. This makes calculating in the metric system simpler, faster, and more error-free.
3. SI is a coherent system. The product or quotient of two unit quantities produces a unit quantity (e.g., m^2—unit of area).

TABLE 5-2. Prefixes and symbols.

Factor	Prefix	Symbol
10^{-12}	pico	p
10^{-9}	nano	n
10^{-6}	micro	μ
10^{-3}	*milli	m
10^{-2}	*centi	c
10^{-1}	deci	d
10^{1}	deka	da
10^{2}	hecto	h
10^{3}	*kilo	k
10^{6}	mega	M
10^{9}	giga	G
10^{12}	tera	T

* most common

4. SI is an absolute system. All base units except the kilogram are defined in terms of physical phenomena.
5. SI is a truly international system.

mass—weight

Mass and weight are used synonymously in everyday life. However, they are not synonymous terms. "Weightlessness" in space brings a clearer picture of the difference between the two terms. Weight can be related to the gravitational pull of the earth, while mass remains the same regardless of location. Weight is a term relating the action of a force (such as gravity) upon a mass. Therefore, most sources indicate that technical work be done in terms of mass, while in general convention and writing mass and weight are used synonymously. In the metric system the unit of measure of force (such as gravity) is the newton. However, for most things convention allows us to measure mass in kilograms.

references

The International (SI) Metric System and How it Works. Robert A. Hopkins. Reseda, California: Polymetric Services, Inc., 1973.

"The Mass—Weight Dilemma" *The Science Teacher.* Volume 39, Number 7, October, 1972. pp. 4, 5.

6

Activities

The activities in this section are designed to give students, teachers, and laymen an understanding of, and a facility in, the use of the metric system. The activities are grouped into the following metric measurement sections.

1. Linear measurement
2. Area and perimeter measurement
3. Mass (weight) measurement
4. Volume and capacity measurement
5. Temperature measurement
6. Culmination activities

Each section includes activities for all levels of understanding. Activities that teach estimation, application, "frame of reference," and general applications are available in each section. In addition, reinforcement activities are also provided. Teachers should select the activities which are the most appropriate for their situation. Students should have available commercial metric materials such as metric rulers, a pan balance, metric weights, a graduated cylinder, a Celsius thermometer, and various capacity containers. But there are ample activities included in this book that teach the metric system without using commercial materials.

The activities can be utilized in many ways. One of the most common is to prepare an activity card for each activity. Then with each activity card the necessary materials can be stored in a box or in some convenient filing system.

The classroom setting can utilize the activities as total-class participation, small groups, or interest center assignments with activity cards.[1] In the primary grades the teacher would need to clarify the directions for the students if students are to work on an individual or small-group basis.

Each activity is conveniently organized to give simple specific information to the user for preactivity preparation and discussion as well as postactivity discussion. The teacher should select the appropriate activity by perusing the teacher objective, the type of activity, and the prerequisites. Make sure that ample time is allowed for each activity.

Each activity consists of the following categories:

Title
Type of activity
Teaching objective
The problem
Number of participants
Materials
Discussion
Prerequisite
Teaching notes
Directions
Extensions

---

[1] Refer to Chapter 1 for details on using centers in the classroom, workshop, or inservice course.

These categories consist of the following:

title. The title is a question which gives an idea of what the activity is trying to accomplish or what the activity is about.

type of activity. The activities have been grouped into three categories. *Exploration* is an activity where no specific result is expected. General familarization with a concept or idea is the expected outcome. *Discovery* activities have an expected outcome and through this discovery it is hoped the metric system will be learned. The *drill and practice* activities include games and written work to reinforce the basic metric concepts, symbols, and vocabulary.

teaching objective. The teaching objective is stated in terms of the student. The student will be able to understand some concept or do some activity. These are stated in general terms and usually the activity includes more specific behaviors than have been indicated in the objective.

the problem. This statement of the problem can be the heading on an activity card or a brief explanation of the problem to be solved, explored, or discussed. The statement of the problem on activity cards or handouts should enhance their use and make it easier for the student to understand his or her responsibilities and objectives.

number of participants. This number gives general guidelines on how many participants could do the activity. Keep in mind that most activities are dependent upon the materials available. In other words, numerous groups could be doing the two or three-participant activities at the same time in a classroom. Written activities are indicated as any number since the teacher prepares the worksheets.

materials. This list indicates all materials which are needed to complete the activity. Substitutions can be made. The most-common materials have been listed. Some of the activities require commercial materials. Homemade substitutes can be utilized in many cases. Do not hesitate to ask the science labs in your school for some of the materials.

discussion. Questions are asked which can be included as preactivity discussion. The questions are ones which usually come up as the student performs the activity. In this way the preactivity discussion can eliminate any insecurity which could develop during the activity. Also some of the questions are excellent for postactivity discussion and the discussion can then be based upon the activity results.

prerequisites. General listing of abilities or understanding required to complete the activities are listed in this category. Keep in mind that these are minimum and oftentimes the desired activity outcomes will not be met due to deficient prerequisite ability on the part of the student. Therefore the student may need some review or concept development to complete the more sophisticated activities.

directions. The step-by-step directions of each activity are included in this category. The directions often refer to teacher-prepared handouts, so be careful to refer to materials and teaching notes before letting the student embark on the activity. If an activity card approach is used, make sure that all materials are coded to the directions so the student understands what materials are to be used to accomplish the objective and complete the activity.

teaching notes. This lists important aspects of the activity of which the teacher should be aware.

extensions. Further ideas for the activity are suggested here. Variations and alternatives of the activity are mentioned. Some of the extensions require materials in addition to the ones mentioned in the materials. These extensions can be for the accelerated or fast-working students or can be used by groups or classes who have not discovered, explored, or understood the concept or objective expected.

Each section also includes a brief introduction of specific skills required or activities which should be noted. The common metric units and symbols are listed plus their decimal and powers-of-10 relationships. At the end of each section are general applications. Teachers should expand on these as students show interest. Refer to metric newsletters, newspapers, and the bibliography for information and metric changes which have already occurred and are occurring daily. The classes should be able to expand on this list monthly.

It is hoped that the reader will feel free to make up his or her own games and activities using the format provided as a guideline. Popular general-classroom activities used in teaching can easily be adapted to metric measurement experiences.

TABLE 6-1. The meter scale

powers of 10						
*kilometer	(km)	=	1000	(10^3)	meters
hectometer	(hm)	=	100	(10^2)	meters
dekameter	(dam)	=	10	(10^1)	meters
*METER	(m)	=	1	(10^0)	meter
decimeter	(dm)	=	0.1	(10^{-1})	meter
*centimeter	(cm)	=	0.01	(10^{-2})	meter
*millimeter	(mm)	=	0.001	(10^{-3})	meter

* Commonly used units

Section I

LINEAR MEASUREMENT

The activities in this section are designed to help the student discover and explore the common linear metric measurements. Drill activities are included for practice. The student will have the opportunity to work with both "standard" and "nonstandard" unit activities as well as to do estimation exercises. Indirect and direct measures are utilized. In several of the activities the student should be able to find some personal standard metric reference for millimeter, centimeter, decimeter, meter, and kilometer. Encourage students to estimate and predict before doing the activity. Opportunities are provided for practice on decimal notation, division and multiplication of decimals, and scientific notation. Students weak in these skills should be referred to Appendix A.

The activity *"how long is the string?"* should be done by all students. This activity will give them an awareness of the metric lengths of their own body. An excellent activity for older students and adults is *"which division is easier?"*. This activity points out the ease of working with metric measurement when compared to the customary system.

Note that several of the activities require teacher-prepared copies for each student. Metersticks or metric rulers are helpful for many of the activities. A chart of the linear measures is helpful for the student if the drill and practice activities are used as part of a learning center.

TABLE 6-2. Linear metric units

Units	Symbol	Multiplication Factors
millimeter	mm	1/1000 m *or* 0.001 m *or* 0.1 cm *or* 10^{-6} km
centimeter	cm	0.01 m *or* 1/100 m
decimeter	dm	100 mm *or* 10 cm *or* 0.1 m *or* .0001 km
meter	m	0.001 km *or* 100 cm *or* 1000 mm
kilometer	km	10^{6} mm *or* 10^{5} cm *or* 1000 m

exploration

Is there a shorter than shortest?
Can you take a giant step?
How long is your ankle?
Does the length change?
How long is the string?
What are your metric measurements?
What is the length of your stride?
How thin is it?

discovery

How thick is it?
How long is a meter?
What are distances like in the out-of-doors?
How long is the tail?
Which path is the longest?
Does the perimeter of a tangram puzzle change?
Do your eyes deceive you?
Which division is easier?
Can you make your globe metric?

drill and practice

What things in your room are a meter in length?
Who can estimate the best?
Can you follow the dots?

Can you draw a straight line?
Which two measures equal the line segment?
How long are the line segments?
Which team is the fastest?
What measures add up to a meter?
Can you unscramble these metric words?
Can you finish the metric puzzle?
How high is it?

general

Applications of linear measures

is there a shorter than shortest?

TEACHING OBJECTIVE

The student will be able to apply correct measuring procedures using nonstandard units.

PROBLEM

Use your own measurement unit to measure different lengths.

NUMBER OF PARTICIPANTS

Any number.

MATERIALS

Soda straws, popsicle sticks, or a measurement unit the class decides upon such as a "Snoopy" or a "Cookie Monster" unit.

DISCUSSION

If you have something to compare an object with, can you find out which object is longer or shorter? How?

PREREQUISITES

None.

DIRECTIONS

1. As a group, decide what you are going to use as a measuring unit.
2. Select two or three objects in different parts of the room which are close to the same length.
3. Use measuring units to rank the objects as "short, shorter, shortest"; "long, longer, longest"; or "high, higher, highest."

4. Use these "units" to find something shorter than the shortest and something longer than the longest.

TEACHING NOTES

The vocabulary development is important, but the concept that something can be shorter than the shortest of a predetermined set is also very important.

EXTENSIONS

Have students measure with a metric ruler those same objects and discuss the role of a standard unit.

can you take a giant step?

TEACHING OBJECTIVE

The student will be able to compare lengths of measure.

PROBLEM

Measure the lengths of your longest and shortest steps.

NUMBER OF PARTICIPANTS

Any number.

MATERIALS

Masking tape, scissors, adding machine tape (strips of paper), pencil.

DISCUSSION

What is the difference between a regular step and a giant step?
Can you take a giant step?

PREREQUISITE SKILLS

None.

DIRECTIONS

1. Place a piece of masking tape on the floor.
2. Have each student put his or her toe on the masking tape and take a giant step.
3. Put a piece of tape where his or her giant step ends.
4. Cut a piece of adding machine tape the length of the student's giant step, and put the student's name on the strip of paper.
5. Once every student has a strip of paper the length of their giant step,

have students order the lengths of their giant steps from longest to shortest.

TEACHING NOTES

An interesting bulletin board can be made by folding all of the strips of adding machine tape in half, ordering them, and taping them on the wall.

EXTENSIONS

Make strips the length of a standing or running broad jump for each student.

how long is your ankle?

TEACHING OBJECTIVE

The student will be able to find lengths equal to a given length.

PROBLEM

Find parts of your body which have the same lengths.

NUMBER OF PARTICIPANTS

Any number.

MATERIALS

String, scissors.

DISCUSSION

How would you measure your waist?
Why wouldn't you use a wooden ruler to measure your waist?

PREREQUISITE SKILLS

None.

DIRECTIONS

1. Cut a piece of string the same length as the circumference of your ankle.
2. Can you find other parts of your body that have the same length as the circumference of your ankle?

EXTENSION

Have students find parts of their bodies that have lengths that are greater than and smaller than the circumference of their ankles.
Have students use another part of their bodies as the basis of comparison.

does the length change?

TEACHING OBJECTIVE

The student will be able to recognize the constancy of length even though units change.

PROBLEM

Use different units of measurement to measure given lengths.

NUMBER OF PARTICIPANTS

Any number.

MATERIALS

Any assortment of objects such as wooden blocks, popsicle sticks, or pieces of string.

DISCUSSION

What is length?

PREREQUISITES

None.

DIRECTIONS

1. Have each group of two or three students select a different unit to measure with.
2. Have the groups all measure and record the width of the door, the length of the chalkboard, and the length of the teacher's desk using their measuring unit.
3. The groups should compare their different results.
4. Discuss the reason for the differences.

TEACHING NOTES

The concept that the length doesn't vary but different results are obtained when different measurement objects are used is important.

EXTENSION

Have groups of students measure various objects using the same measurement unit and then discuss any discrepancies that arise.

how long is the string?

TEACHING OBJECTIVE

The student will be able to measure using standard and nonstandard units.

PROBLEM

Estimate lengths using string. Then verify your measurements.

NUMBER OF PARTICIPANTS

Any number.

MATERIALS

String, scissors, centimeter cubes or strips.

DISCUSSIONS

What is the value of a standard unit of measure?
What does it mean to estimate?

PREREQUISITES

None.

DIRECTIONS

1. Measure each of the items listed below using your string as a measuring stick. Cut the string when you have found the correct measures.
   a. length of your little finger
   b. distance around your fist
   c. distance around your waist
   d. distance around your wrist
   e. distance around your thumb
   f. length of your math book

g. distance around the handle of a baseball bat

h. distance around a door knob.

2. Estimate the length of each piece of string.

3. Now measure the length of your strings using centimeter cubes or strips.

TEACHING NOTES

Discuss with the students the need for standard units when communicating with each other.

EXTENSIONS

Have students compare measures to determine ratio, e.g., circumference of waist to circumference of neck.

what are your metric measurements?

TEACHING OBJECTIVE

The student will be able to determine the length of various parts of his or her body using appropriate metric measures.

PROBLEM

Find the length of various parts of your body, e.g., hand span, reach, height, width of little finger, etc.

NUMBER OF PARTICIPANTS

Any number.

MATERIALS

String, scissors, metric ruler, or metric tape measure.

DISCUSSION

What would be the value of knowing your measurements? Could you use this knowledge to measure other objects? How?

PREREQUISITES

Ability to estimate.
Ability to read and use a metric ruler or metric tape measure.

DIRECTIONS

1. Decide on an appropriate unit of measure.
2. Estimate and then measure each of the following:
   a. hand span (fingers spread wide apart)
   b. reach (both arms outstretched)
   c. height
   d. width of little finger

e. circumference of neck
f. length from elbow to end of longest finger
g. length from nose to end of outstretched hand

TEACHING NOTES

If a metric ruler is used, have children measure the parts of their bodies using string and then have them measure the string to determine body measurements.

EXTENSION

Have the students measure various objects in the room utilizing their body measurements. Then have them use the metric ruler to actually measure the objects. How do these measurements compare?

what is the length of your stride?

TEACHING OBJECTIVE

The student will be able to measure the length of his or her average stride.

PROBLEM

Find the average length of your stride and use this information to find the approximate length of your classroom.

NUMBER OF PARTICIPANTS

Any number.

MATERIALS NEEDED

Masking tape or chalk, metric ruler and/or trundle wheel.

DISCUSSION

What would be the value of knowing the length of your stride?
In what types of situations would you be most likely to use the length of your stride as a unit of measure?

PREREQUISITES

Ability to read and use a metric ruler or trundle wheel. Ability to divide by 10.

DIRECTIONS

1. On the floor place a small piece of tape, or make a mark with chalk.
2. Starting at the tape or chalk mark, take ten normal walking steps.
3. Put another piece of tape, or chalk mark, where your last step ended.
4. Using either the trundle wheel or metric ruler, measure the distance between the two marks on the floor.

5. Find the average length of your stride.

6. Using your average stride as a measure, find the approximate length of the classroom. Use the trundle wheel or metric ruler to check your answer.

TEACHING NOTES

If space permits, have students find their average stride by taking 100 normal walking steps instead of 10.

EXTENSIONS

Once students have determined the length of their stride, have them check their walking measurement against the distances measured with the metric ruler for each of the following distances:

1. distance around the classroom
2. distance around the outside of the school
3. distance between telephone poles
4. distance around basepaths on ball diamond
5. distance around playground
6. distance to the principal's office
7. distance around the circle formed when all of the students hold hands and stretch out as far as they can
8. a guessed distance of 25 m, 50 m, and 100 m

how thin is it?

TEACHING OBJECTIVE

The student will be able to measure in millimeters.

PROBLEM

Find the thickness of a playing card.

NUMBER OF PARTICIPANTS

Any number.

MATERIALS

Micrometer and/or metric ruler and ten cards for each student. (*Note:* Playing cards, index cards, 5 cm X 10 cm poster boards, etc. can be used.)

PREREQUISITES

Ability to use and read to the nearest millimeter the micrometer and/or metric ruler.

DIRECTIONS

1. Using your micrometer and/or metric ruler find the thickness of ten cards. Use millimeters as your unit of measure.
2. Once you have determined the thickness of ten cards, determine what the thickness of one card would be without measuring.

TEACHING NOTES

Stress the following types of relationships to the students:

75 cm = 0.75 m

75 mm = 0.075 m

75 mm = 7.5 cm

If the students need practice, have them do some practice exercises showing the moving of the decimal point when you are dividing by 10.

EXTENSIONS

Find the thickness of one page in a book using the above procedure. The student may need to measure more than ten pages.

how thick is it?

TEACHING OBJECTIVE

Student will be able to measure using metric calipers.

PROBLEM

Using vernier calipers, measure parts of your body and various parts of your environment.

NUMBER OF PARTICIPANTS

Any number.

MATERIALS

Large vernier calipers.

DISCUSSION

What types of objects would you measure with vernier calipers?

PREREQUISITES

Ability to read and use a vernier caliper.

DIRECTIONS

1. Using the vernier calipers, estimate and then measure the following items in the classroom:
   a. width of your shoulders
   b. width of your head
   c. thickness of five dictionaries
   d. thickness of table
   e. diameter of a soccer ball
2. Using the vernier calipers, estimate and then measure the following items found outside:

a. diameter of tree
b. width of a stop sign
c. width of bicycle tire
d. thickness of the flag pole

TEACHING NOTES

A workable vernier caliper can be made by using two T-squares or carpenter squares and a metric ruler as in Figure 2.

EXTENSIONS

Many other objects can be measured besides those mentioned.

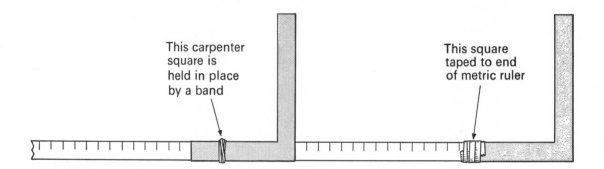

FIGURE 2. Vernier caliper.

how long is a meter?

TEACHING OBJECTIVE

The student will use a standard unit (meter) for measuring distances.

PROBLEM

Use the meter measurement and find all the things in the room with a length of 1 m.

NUMBER OF PARTICIPANTS

Any number.

MATERIALS

Metric ruler or strip of adding machine paper that has been pre-measured to 1 m for each student.

DISCUSSION

Why would it be important for each person to have the same length measuring device when we are measuring the same objects?

PREREQUISITE SKILLS

None.

DIRECTIONS

1. Using your metric ruler (strip of paper) find all the things that are about 1 m long in your room.
2. How many things can you find that are just a little longer than 1 m?
3. How many things can you find that are just a little shorter than 1 m?

TEACHING NOTES

This activity can lead naturally into a discussion of multiples and divisions of a unit length.

EXTENSION

Have students find objects that are equal to, longer than, or shorter than 0.5 m.

what are distances like in the out-of-doors?

TEACHING OBJECTIVES

The student will be able to measure distance using the trundle wheel.

PROBLEM

Use your trundle wheel to determine distances.

NUMBER OF PARTICIPANTS

Any number.

MATERIALS

Trundle wheel.

DISCUSSION

How do you use a trundle wheel?

PREREQUISITES

Ability to measure with a trundle wheel.

DIRECTIONS

1. Using a trundle wheel, measure and record the following:
   a. distance around the classroom
   b. distance around the outside of the school building
   c. distance between telephone or electrical poles
   d. distance around basepaths on ball diamond
   e. distance around playground
   f. distance to the principal's office
   g. distance around the closed figure formed when all of the students hold hands and stretch out as far as they can

2. Estimate a distance of 10 m, 50 m, and 100 m. Using a trundle wheel, check the accuracy of your estimates.

TEACHING NOTES

Be sure students are aware of the unit of length which the trundle wheel measures. Preliminary estimating of the circumference of the trundle wheel would be helpful.

EXTENSIONS

Have students use the trundle wheel to measure a kilometer.

how long is the tail?

TEACHING OBJECTIVE

The student will learn to use a metric ruler to measure in centimeters.

PROBLEM

Determine the metric measure of the tails of the animals.

NUMBER OF PARTICIPANTS

Any number.

MATERIALS

Set of animals (monkeys, cats, oppossum, etc.) with different lengths string for tails, metric ruler. (See Figure 3.)

FIGURE 3. Animal figures.

DISCUSSION

Are the tails of the animals you have all the same size?

Suppose you could not move the animals, how could you find out which animal had the longest tail?

PREREQUISITE SKILLS

None.

DIRECTIONS

Order your animals from the animal with the longest tail to the animal with the shortest tail.

TEACHING NOTES

This would be a good exercise to introduce the reading of centimeters on the metric ruler.

EXTENSION

Have students measure the tail of each animal. Other animal pictures could be used, with string in lieu of part of their body. The bodies of various animals could be measured.

which path is the longest?

TEACHING OBJECTIVE

The student will be able to accurately measure a small distance.

PROBLEM

Can you draw any of the given figures in one continuous line and then accurately estimate the length of the line segment?

NUMBER OF PARTICIPANTS

Any number.

MATERIALS

Prepared worksheet and metric ruler.

DISCUSSION

By glancing at these figures, which one do you think can be drawn without lifting your pencil or retracing?

PREREQUISITES

Ability to read and use the metric ruler.

DIRECTIONS

1. Look at the four figures drawn in Figure 4.
2. Some of these figures can be made by drawing one continuous line without lifting the pencil off the paper and without retracing any line.
3. Estimate the length of the line segment of each figure if it were possible to connect the lines into one continuous line segment.
4. See if you can trace each figure without lifting your pencil off the paper and without tracing over any line.

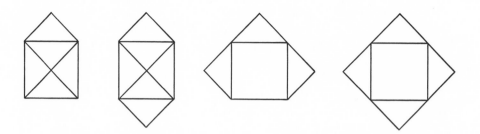

FIGURE 4. Which are continuous line segments?

5. For those you can draw with one continuous line, measure with your metric ruler the length of all line segments.
6. Compare your actual measurement with your estimate.

TEACHING NOTES

Have students determine the minimum number of measures required to find the total length of each figure.

EXTENSIONS

Make other figures (networks) for students to measure.

does the perimeter of a tangram puzzle change?

TEACHING OBJECTIVE

The student will be able to accurately measure using millimeters.

PROBLEM

Find the perimeter of a tangram puzzle and determine if the perimeter of the puzzle changes as you change the shape of the figure using all pieces of the tangram puzzle.

NUMBER OF PARTICIPANTS

Any number.

MATERIALS

Prepared worksheet and metric ruler.

DISCUSSION

What is a tangram?
Is there only one polygon that can be formed with the seven tangram pieces?

PREREQUISITES

Ability to read and use the metric ruler.
Understanding of perimeter measure.

DIRECTIONS

1. Estimate the distance around the outside of each of the designs below (perimeter).
2. Using your metric ruler, measure the perimeter of the square formed with the tangram pieces.

3. Cut out the tangram pieces.
4. Form each of the designs in Figure 5.
5. As you form each design, measure the perimeter of the design using your metric ruler.
6. Was the perimeter the same for each figure?
7. What was the difference between your estimate and the actual measurement?

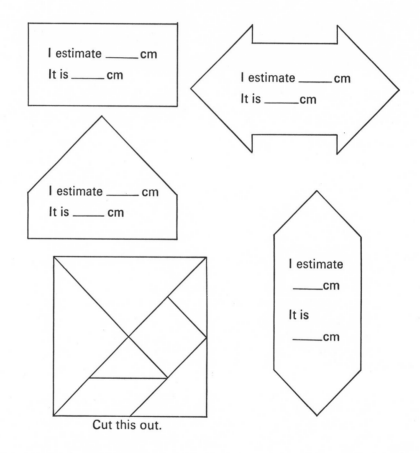

I estimate _____ cm

It is _____ cm

I estimate _____ cm

It is _____ cm

I estimate _____ cm

It is _____ cm

I estimate _____ cm

It is _____ cm

Cut this out.

FIGURE 5. Estimate the perimeters.

TEACHING NOTES

Discuss the findings of questions 6 and 7 with the class once the students have finished doing the exercises.

EXTENSIONS

Have students form other figures using the tangram pieces and have them measure the perimeter of the new figures they have formed. Are they different? Why?

do your eyes deceive you?

TEACHER OBJECTIVE

The student will estimate and then measure line segments.

PROBLEM

Find out how accurately you can estimate the length of a line segment.

NUMBER OF PARTICIPANTS

Any number.

MATERIALS

String, metric ruler, prepared worksheet.

DISCUSSION

How could we measure something that isn't a straight line?

PREREQUISITES

Ability to read and use a metric ruler.

DIRECTIONS

1. Look at the line segments in Figure 6.
2 Estimate which figure or line segment is the longest.
3. Measure each of the figures or line segments to see how accurate your estimates were.

TEACHING NOTES

Stress that the measuring devices are used to communicate measurement.

FIGURE 6. Estimate, then measure.

EXTENSION

Have students create their own optical illusions. Make sure that all measurements are done in metric.

which division is easier?

TEACHING OBJECTIVES

The student will be able to list advantages of the metric system over the customary system of measuring.

PROBLEM

By working a similar problem in the customary system and the metric system, decide which system is easier to use.

NUMBER OF PARTICIPANTS

Any number.

MATERIALS

Paper, inch/foot ruler, metric ruler, scissors, and pencil.

DISCUSSION

Which method of measuring (customary or metric) do you feel more comfortable with? Why?

Which method of measuring do you feel is easier to work with? Why?

Do you feel one system of measuring is superior to the other? Which system? Why?

PREREQUISITES

Ability to use and read inch/foot ruler.
Ability to use and read metric ruler.
Understanding of how to divide a line segment into equal parts.
Ability to divide by 5.
Understanding the relationship between centimeters and millimeters.

DIRECTIONS

1. Using your ruler, measure a strip of paper that is 12 inches long. ~~You may have to measure on the diagonal of your paper.~~

2. ~~Using your metric ruler,~~ *and* measure a strip of paper that is 30.5 cm long. ~~You may have to measure on the diagonal of your paper.~~

3. Compare the strips.

4. Do you think it will be easier to divide the strip of paper into five equal parts using your inch/foot ruler or using your metric ruler to help you?

5. Using your inch/foot ruler, divide the 12-inch strip of paper into five equal parts.

6. Using your metric ruler, divide the 30.5-cm strip into five equal parts.

7. Which was easier to do?

TEACHING NOTES

Point out that 30.5 cm is the same as 305 mm, which is easily divisible by 5. Discuss why the metric computations were easier.

EXTENSION

Choose other lengths that have the same value in the customary and metric systems and try dividing them into equal parts.

can you make your globe metric?

TEACHING OBJECTIVE

The student will be able to make metric scale drawings.

PROBLEM

Using scale measurement, find the distance from Phoenix to various parts of the world on your globe.

NUMBER OF PARTICIPANTS

Groups of three.

MATERIALS

Globe, string, and metric ruler (or metric tape measure in place of string and metric ruler).

DISCUSSION

Are distances found on the globe exact or approximate?
Would the error of measurement be greater or less on a city map? Why?
What is a scale drawing? When do we use them?

PREREQUISITES

Familiarity with globe.
Ability to read and use a metric ruler or metric tape measure.
Ability to perform division operation.
Understanding of scale measurement.

DIRECTIONS

1. Using the string, measure the equator of your globe.

2. The length of the equator is approximately 40 000 km.
3. The scale of your globe becomes 1 cm = _____ km.
4. Use this scale to find the following distances:
   a. Phoenix to New York
   b. Phoenix to London
   c. Phoenix to Capetown
   d. Phoenix to Mexico City
   e. Phoenix to Tokyo
   f. Phoenix to Moscow

TEACHING NOTES

Help students find the scale measure by taking the length of the string and dividing this number into 40 000 km.

EXTENSIONS

Have students find other distances by replacing Phoenix with their home city.

Have students exchange globes and do the same discovery exercises again to see if using different-sized globes would produce different scales.

what things in your room are a meter
in length?

TEACHING OBJECTIVE

The student will be able to estimate the length of a meter.

PROBLEM

Estimate items in your classroom that are 1 m in length and then verify your findings by measuring.

NUMBER OF PARTICIPANTS

Any number.

MATERIALS

Metric ruler (or some object 1 m long, such as a piece of string or a strip of adding machine tape cut to that length).

DISCUSSION

How long is a meter?
What does it mean to estimate?

PREREQUISITES

None.

DIRECTIONS

1. Have each student demonstrate with their hands the length of a meter.
2. Have each student compare their estimate to a meter measure.
3. Using their metric measure have students:
   a. find several objects in the classroom that are about 1 m in length.

b. find points on their bodies that are 1 m apart.
c. make a list of ten things they think are about 1 m in length, then measure the items and sort them into groups that are longer, shorter, or equal to 1 m in length.
d. draw a line on the chalkboard to show what they think is the length of a meter. Now measure the line to determine the accuracy of their estimate.

TEACHING NOTES

Remember, practice is needed to help students think metric.

EXTENSIONS

Have students take their meter measure home and find items that are about 1 m in length.

who can estimate the best?

TEACHING OBJECTIVE

The student will be able to estimate the length of line segments.

PROBLEM

Determine which array of 12 dots would yield the longest path and the shortest path if you connected all 12 dots with straight lines.

NUMBER OF PARTICIPANTS

Any number.

MATERIALS

Centimeter dot paper and metric ruler.

PREREQUISITES

Ability to read and use a metric ruler.

DISCUSSION

What is an array?
How would you determine which is the longest or shortest path?

DIRECTIONS

1. On your centimeter dot paper make as many different rectangular arrays of 12 dots as you can.
2. For each array estimate what would be the length of the shortest possible path to connect all 12 dots. No lines should intersect.
3. Which array would have the shortest path?
4. Which array would have the longest path?

5. Now connect the dots, measure, and see if your estimates were correct.

TEACHING NOTES

Have students measure the length of the diagonal lines and compare to horizontal and vertical line segments.

EXTENSIONS

Try this activity with triangular arrangements.

can you follow the dots?

TEACHING OBJECTIVE

The student will be able to measure accurately metric lengths in centimeters.

PROBLEM

Find the hidden picture by following carefully the given directions.

NUMBER OF PARTICIPANTS

Any number.

MATERIALS

Copy of the diagram in Figure 7 (completed drawing for teacher's use only), metric ruler, and pencil.

PREREQUISITES

Ability to estimate centimeters.
Ability to read and use a metric ruler.
Ability to do dot-to-dot puzzles.

DIRECTIONS

1. Starting at point *A;* find a point that is 6 cm away and draw a line from point *A* to that dot.
2. Now from this point find a point that is 2.5 cm away and draw a straight line to that dot.
3. From this point draw a straight line to another point that is about 2.5 cm away.
4. From this point draw a straight line to a point that is about 6 cm away.
5. Next draw a line from this point to a point that is 1.25 cm away.

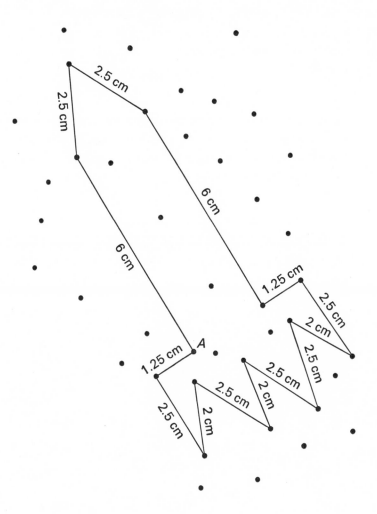

FIGURE 7. Connect the dots.

6. Now draw a line from this point to a point that is 2.5 away.
7. From here draw a line to a point that is 2 cm away.
8. Now draw a line from this point to a point that is 2.5 cm away.
9. From here draw a line to another point that is 2.5 cm away.
10. Next draw a line from this point to a point that is 2 cm away.
11. Now draw a line from this point to another point that is 2.5 cm away.
12. Draw a line from this point to a point that is 2 cm away.

13. Again draw a line from this point to a point that is 2.5 cm away.

14. To complete the picture, draw a line from this point to point *A*. This line segment should be 1.25 cm in length.

15. Is the picture you drew a rocket?

TEACHING NOTES

If students have never done any dot to dot drawings, it would be worthwhile to help them do two or three moves to get them started. Remind them that the line is continued from where the previous line ended.

EXTENSIONS

Have students make up some dot-to-dot puzzles utilizing metric measures and have their fellow students solve them.

can you draw a straight line? [1]

TEACHING OBJECTIVE

The students will be able to draw line segments accurately to the nearest centimeter.

PROBLEM

Win the game by strategically constructing the line segments shown on the cards you draw. First to Square 4 wins.

NUMBER OF PARTICIPANTS

Any number.

MATERIALS

Game sheet (Figure 8) for each player, stack of prepared cards, and metric ruler.

PREREQUISITES

Ability to read and use a metric ruler.

DIRECTIONS

Prior to playing this game, prepare the following cards. (*Note:* If blank cards are not available, cut poster board 5 cm by 10 cm.) Place one measurement to a card.

0.2 cm	4.2 cm	8 cm
0.5 cm	4.5 cm	8.2 cm
1 cm (2)	4.8 cm	8.6 cm
1.5 cm	5 cm (2)	9 cm
1.8 cm	5.2 cm	9.3 cm

[1] Adapted from Linus' Lines developed as part of *Project Colamda Title III* ESEA Castle Rock, Colorado.

FIGURE 8. Game sheet.

2 cm	5.5 cm	9.6 cm
2.2 cm	5.9 cm	10 cm
2.5 cm	6 cm	10.2 cm
2.8 cm	6.3 cm	10.5 cm
3 cm (2)	6.6 cm	11 cm
3.3 cm	6.8 cm	11.6 cm
3.5 cm	7 cm	11.8 cm
3.8 cm	7.2 cm	12 cm
4 cm (2)	7.4 cm	12.4 cm
13 cm	13.6 cm	14.4 cm

1. Shuffle the cards and place them face down on the table.
2. The first player draws a card and on his or her game sheet constructs a line segment equal to the length indicated on the card. The first segment must start from the sign "Metric" going in any direction.
3. Each player proceeds as the first player did using their own game board.
4. On each player's next turn another card is drawn. The player continues the line from where the previous line ended. The line segment may go in any direction.
5. The object of the game is for a line segment to end within Square 1, then continue in segments to Square 2, then 3, and then 4.
6. The first player to get through Squares 1, 2, 3, and within Square 4 with line segments wins the game.

Note: A line segment may not pass completely through any square during the game. Segments may intersect.

which two measures equal the line segment?

TEACHING OBJECTIVE

The student will be able to accurately estimate the length of line segments.

PROBLEM

Estimate length of line segments and verify your findings by measuring.

NUMBER OF PARTICIPANTS

Any number.

MATERIALS

Prepared worksheet and metric ruler.

DISCUSSION

When is it important to be able to estimate line segments?
What are some techniques that you could use to help you estimate?
Are there some common objects that you could use to help you estimate? What are they?

PREREQUISITES

Ability to estimate.
Ability to read and use a metric ruler.
Understanding of the relationship between centimeter and millimeter.

DIRECTIONS

1. Estimate the length of each line segment in Figure 9.

7 cm	4 cm	2 cm	1 cm
15 mm	35 mm	6 cm	7 cm
8 cm	12 cm	1 cm	3 cm
9 cm	35 mm	3 cm	55 mm
5 mm	8 cm	25 mm	9 cm
11 mm	2 cm	14 mm	3 cm

FIGURE 9. Estimate, then measure.

2. Circle the measures under the segment which you think will equal the length of the line segment when the measures are added together.
3. When you have finished the page, check your answers with a metric ruler to see if you were correct.

TEACHING NOTES

Have students decide when the use of the centimeter and millimeter measuring units are appropriate.

EXTENSIONS

Have students estimate line segments drawn on the board.

Do an exercise similar to the one on this page, but this time have the students add three metric measures to equal a given line segment, using centimeters.

how long are the line segments?

TEACHING OBJECTIVE

The student will be able to measure line segments to the nearest centimeter and millimeter.

PROBLEM

Find the length of the given line segments.

NUMBER OF PARTICIPANTS

Any number.

MATERIALS

Prepared worksheet and metric ruler.

PREREQUISITES

Ability to read and use a metric ruler.
Understanding of the relationships between metric measurements.

DIRECTIONS

Using a metric ruler, measure the line segments in Figure 10.

1. $\overline{AB}$ = _____ cm or _____ mm
2. $\overline{CD}$ = _____ cm or _____ mm
3. $\overline{CE}$ = _____ cm or _____ mm
4. $\overline{CF}$ = _____ cm or _____ mm
5. $\overline{AC}$ = _____ cm or _____ mm
6. $\overline{BD}$ = _____ cm or _____ mm
7. $\overline{CD} - \overline{EF}$ = _____ cm or _____ mm
8. $\overline{AB} + \overline{BF} + \overline{EF} + \overline{AE}$ = _____ cm or _____ mm
9. $\overline{BD} + \overline{BF} - \overline{FD}$ = _____ cm or _____ mm
10. $\overline{AC} + \overline{AE} + \overline{CE} - \overline{EF}$ = _____ cm or _____ mm

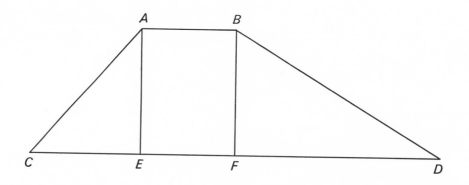

FIGURE 10. Measure the segments.

TEACHING NOTES

Remind the students of the following relationships:

1 cm = 10 mm

1 m = 100 cm

EXTENSIONS

Have students write answers in millimeters, centimeters, decimeters, and meters to see if they can change easily from one unit to another.
Have students make up their own figures and measure them.

which team is the fastest?

TEACHING OBJECTIVE

The student will be able to measure distances accurately and quickly.

PROBLEM

Be the winning team by measuring accurately a given set of objects before any other team can complete the task.

NUMBER OF PARTICIPANTS

Any number.

MATERIALS

Paper, pencil, and metric ruler.

DISCUSSION

When is accuracy important?
How accurate should the measures be?

PREREQUISITES

Ability to read and use a metric ruler.

DIRECTIONS

1. Divide the class into teams of equal ability.
2. List on the board at least ten items in the room whose distance can be determined. (Dittoed list could be prepared ahead of time.)
3. Have each team copy the list from the board.
4. At a given signal the first member of the team measures the first item on

the list and then passes the list to the second player who measures the second item on the list, etc.

5. The first team to finish measuring all of the items on the list correctly is the winner.

TEACHING NOTES

Decide in advance the accuracy that is acceptable.

EXTENSIONS

Have students change all measurements given to meters, centimeters, and millimeters.

Have each student add the length of his or her item on the list to the sum of the previous items for each team. The team with the most accurate sum is the winner.

what measures add up to a meter?

TEACHING OBJECTIVE

The student will be able to change and combine metric measures.

PROBLEM

Add the measures, circle those that add up to 1 m. Find more than your classmates and be the winner.

NUMBER OF PARTICIPANTS

Any number.

MATERIALS

Prepared worksheet.

DISCUSSION

How does conversion within the metric system compare to conversion within the customary system?

PREREQUISITES

Understanding of relationships between centimeter, meter, decimeter, and millimeter.
Ability to add linear measures.

DIRECTIONS

Figure 11 has 36 squares. Connect any two or three squares that touch at any point so that the sum is 1 m.

TEACHING NOTES

This could be done as a competitive game by having students initial and count all of the squares they enclose.

100 mm	60 cm	80 cm	6 dm	100 mm	60 cm
9 dm	10 cm	100 mm	200 mm	200 mm	30 cm
70 cm	1 dm	800 mm	20 cm	2 dm	600 mm
40 cm	30 cm	300 mm	5 dm	100 mm	3 dm
5 dm	200 mm	300 mm	40 m	60 cm	7 dm
10 cm	8 dm	30 mm	400 mm	300 m	20 cm

FIGURE 11. Connect the appropriate squares.

EXTENSIONS

A more difficult version of this activity could be presented depending on mathematical ability of students (see Figure 12).

30.5 cm	695 mm	7.4 cm	6.3 dm	0.7 m	200 mm
515 mm	3 dm	18.5 cm	3.7 dm	10 cm	0.1 m
1.8 dm	400 mm	32 cm	0.1 m	0.8 m	5.4 cm
82 cm	0.0005 km	0.4 m	90 cm	0.25 m	46 mm
7.3 cm	10 cm	416 mm	36 cm	75 cm	0.9 m
2.7 cm	0.9 m	58.4 cm	640 mm	56 cm	0.44 m

FIGURE 12. Connect the appropriate squares.

can you unscramble these metric words?

TEACHING OBJECTIVE

The student will be able to recognize metric vocabulary.

PROBLEM

Unscramble the letters to form words having something to do with linear measurement.

NUMBER OF PARTICIPANTS

Any number.

MATERIALS

Each student will need a copy of the scrambled words.

DISCUSSION

What words do we use when measuring linear distances?

PREREQUISITES

Ability to spell linear measuring words.

DIRECTIONS

1. Unscramble the following words. Each word has something to do with linear measurement.
2. Check to see that all words are spelled correctly.
   a. trim ce   *metric*
   b. near li   *linear*
   c. gen lh t   *length*
   d. is t dance   *distance*
   e. ent   *ten*

f. lime lim ter   *millimeter*
g. i met net rec   *centimeter*
h. er til moke   *kilometer*
i. tem er   *meter*
j. ruse a me   *measure*

TEACHING NOTES

Discuss System International (SI) rules on metric vocabulary and symbols.

EXTENSIONS

Have a race to see who can unscramble the listed words in the least amount of time.

Have students develop their own scrambled word lists.

can you finish the metric puzzle?

TEACHING OBJECTIVE

The student will be able to use metric vocabulary when appropriate.

PROBLEM

Do the following crossword puzzle.

NUMBER OF PARTICIPANTS

Any number.

MATERIALS

Each player will need a copy of the puzzle in Figure 13.

DISCUSSION

What is the purpose of doing crossword puzzles?

PREREQUISITES

Understanding of how to do a crossword puzzle.
Knowledge of the metric linear vocabulary and symbols.

DIRECTIONS

1. Read each of the clue statements given under the puzzle.
2. When you think you know the proper word, write the word in the appropriate blanks of the puzzle.

TEACHING NOTES

Discuss the rules and symbols of the metric system as outlined by System International (SI).

EXTENSION

Have students make up their own crossword puzzles utilizing metric words.

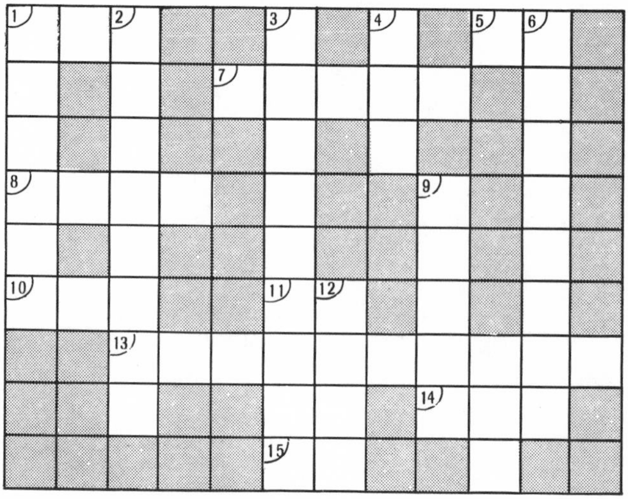

Across

1. Another name for a boy
5. Symbol for kilometer
7. A _____ meter is 1/1000 of a meter.
8. Door on a fence
10. An animal that lays eggs
11. Nickname for Edward
13. 1/100 of a meter
14. A bright color
15. Note of the musical scale

Down

1. _____ X width = area
2. We use the meter to measure _____.
3. 1000 meters
4. Sick
6. To find out how long something is, it is _____.
9. A base unit in the metric system
12. 10¢

FIGURE 13. A crossword puzzle.

how high is it?

TEACHING OBJECTIVE

The student will be able to calculate using metric measurements.

PROBLEM

Find out how thick various stacks of money would be.

NUMBER OF PARTICIPANTS

Any number.

MATERIALS

None.

DISCUSSION

Where would you find information concerning the federal budget?

PREREQUISITES

Ability to multiply by powers of 10.

DIRECTIONS

Given that a stack of 100 one-dollar bills is 9.5 mm high, what would be the height of the following stacks?

1. $1000
2. $1 000 000
3. $1 000 000 000
4. this year's federal budget

TEACHING NOTES

Be sure students use correct factors. Note the ease of using the other metric measures.

EXTENSIONS

Suppose that the stack of 100 one-dollar bills is 8.6 mm high. What would be the height of the stacks? Express answers in scientific notation.

fill in the blanks

1 Speed limits will change!
    a. A 50 miles per hour speed limit will become an 80 kilometers per hour speed limit.
    b. A 15 mph speed limit will become a _____ kmph speed limit.
    c. A 35 mph speed limit will become a _____ kmph speed limit.
2. Distance signs between cities will change!
    a. If a sign says "Flagstaff 10 miles" it will become "Flagstaff 16 kilometers."
    b. If a sign says "Denver 100 miles" it will become "Denver _____ km.
    c. If a sign says "Spokane 30 miles" it will become "Spokane _____ km."
3. Sports measures will change?
    a. What would be an appropriate unit for measuring the high jump metrically? _____
    b. In metric units what would be the approximate world metric high jump record? _____
    c. The mile-run distance is approximately _____.
    d. What is the approximate metric measure of a football field? _____
    e. How high is a basketball goal in metric units? _____
4. What other linear measures will change?

Section II

AREA AND PERIMETER

Activities for both area and perimeter are included in this section. Many of the activities require predicting, guessing, and estimating. Encourage students to estimate what they think the answer will be and to predict what they think will happen *before* they do the activity. The exploration activities vary from the general to the specific.

The geoboard is an excellent manipulative for introducing the concept of perimeter. It is important for the student to understand the concept of perimeter. Some of the general activities of finding the "distance around" require using string and then measuring the length of the string to determine the object's perimeter.

The geoboard is also an excellent device to illustrate and to help students understand the concept of area. To begin area activities the student needs to have a concept of a unit to measure area. This could be a square unit, triangular unit, etc. Students need to be aware that the notation for square meter is m^2. Since this may be their first experience with the exponent, a review of Appendix A may prove helpful. Point out to the students that when working with area units the decimal point moves in 100's instead of 10's. A helpful hint is that the exponent "2" requires a change of two places for each unit change. Extra drill similar to *"where is the decimal point?"* is encouraged.

TABLE 6-3. Perimeter and area metric units

Units	Symbol	Multiplication Factors
Perimeter		
millimeter	mm	0.001 m *or* 1/1000 m
centimeter	cm	0.01 m *or* 1/100 m
decimeter	dm	0.1 m *or* 1/10 m *or* 10 cm *or* 100 mm
meter	m	0.001 km *or* 100 cm *or* 1000 mm
kilometer	km	10^6 mm *or* 1000 m
Area		
square millimeter	mm²	0.01 cm² *or* 1/100 cm²
square centimeter	cm²	0.000 1 m² *or* 1/10 000 m²
square meter	m²	10^6 mm² or 10^4 cm²
hectare	ha	10^4 m²
square kilometer	km²	100 ha *or* 10^6 m²

The hectare, which replaces the acre unit in land measurement, does not follow the general prefix format but does follow the power-of-10 notation.

It would be helpful to provide a chart of the measures and decimal notations used in area and perimeter for the beginning student.

exploration

What shape are you?
Can you guess the number?
Does the form make a difference?
What if you had a square shape?
Can you match the area?
What is the area and perimeter of your silhouette?
Could you carpet your school halls?
Are sports metric?
What is the shape of your school ground?

discovery

How do you predict?
Can you make the metric squares?

Does the surface area change?
Which gives more—a square or a rectangle?
Where did the area go?
How long is the path?

drill and practice

Are 20 questions enough?
Can you close the figure?
Where is the decimal point?
Can you guess?
Which triangles fit?
Which is easier?

general

Applications of area and perimeter measures

which shape are you?

TEACHING OBJECTIVE

The student will be able to compare his or her outstretched arms' length and his or her height using string.

PROBLEM

Determine whether your arm reach is longer or shorter than your height and the shape you form.

NUMBER OF PARTICIPANTS

Any number, students will work in pairs.

MATERIALS

String, scissors.

DISCUSSION

Which do you think is longer: your height, or your reach when your hands are spread wide apart?

PREREQUISITE SKILLS

Ability to differentiate between a square and a rectangle.

DIRECTIONS

1. Spread your arms wide apart and cut two pieces of string the same length as your reach.
2. Cut two pieces of string the same length as your height.
3. Compare the length of your reach with the length of your height.
4. If the lengths are the same form a square with the four pieces of string.

5. If the lengths of the string are not the same, then form a rectangle on the floor with the four pieces of string. The string length of your reach should form the top and bottom of your rectangle and the string length of your height should form the sides of your rectangle.
6. Is the rectangle you formed a tall rectangle or a wide rectangle?
7. What shape are you?

TEACHING NOTES

Students should have adequate room to lay these out so that comparisons can be made.

EXTENSIONS

Determine what portion of the class form squares, tall rectangles, and wide rectangles. Compare the results of your class with another class in the building. Is there a difference between the shape of the boys and girls in the class? Graph your results.

can you guess the number?

TEACHING OBJECTIVE

The student will be able to predict the number of units needed to cover a certain area.

PROBLEM

Determine the number of units needed to cover each area.

NUMBER OF PARTICIPANTS

Any number.

MATERIALS NEEDED

Centimeter cubes or blocks or squares such as wooden blocks.

DISCUSSION

How would we measure a surface?

PREREQUISITE SKILLS

Ability to count.
Recognition of square and rectangle.

DIRECTIONS

1. Make an outline of a square or rectangle using the centimeter cubes or blocks.
2. Predict the number of centimeter cubes or blocks it would take to fill the inside.
3. Fill the inside of the figure. How many blocks did it take to fill the inside?

4. Determine the number of blocks used in all.
5. Repeat for a wide variety of sizes and shapes of rectangles.

TEACHING NOTES

Colored rods can be very beneficial here if the student is quite familiar with them. Make perimeters with different rods and the student will relate a color with a number and likely eliminate a lot of counting.

EXTENSION

Have students form a square and predict the number of units needed for a border.

does the form make a difference?

TEACHING OBJECTIVE

The student will demonstrate that the perimeter remains constant even though the shape of a figure changes.

PROBLEM

Using a trundle wheel determine if the perimeter remains constant when the shape of the figure changes.

NUMBER OF PARTICIPANTS

All class members.

MATERIALS

Trundle wheel.

DISCUSSION

As you change figures, what will happen to the distance around the outside?

PREREQUISITE SKILLS

None.

DIRECTIONS

1. Students will clasp hands to form a large circle. They are to stretch out as far as they can.
2. With the trundle wheel determine the distance around the outside.
3. Determine if the distance changes when the students form:
   a. a rectangle of various lengths and widths

b. a square
c. a triangle
d. an oval

TEACHING NOTES

This activity can help students get a good understanding of perimeter and gives them practice in measuring with a trundle wheel.

EXTENSIONS

Repeat the same activity using string.

what if you had a square shape?

TEACHING OBJECTIVE

The student will be able to calculate the area of a rectangular surface.

PROBLEM

Determine the area you would occupy if you were rectangular in shape.

NUMBER OF PARTICIPANTS

Any number.

MATERIALS

Metric ruler and masking tape.

DISCUSSION

Why is the square the easiest surface area to determine?

PREREQUISITE SKILLS

Ability to understand the concept of area.
Ability to multiply.

DIRECTIONS

1. Stand against a wall with your arms outstretched.
2. Have a friend place a small piece of masking tape to mark the top of your head and the end of each of your outstretched hands.
3. Measure from the floor to the tape above your head. It was how many centimeters?
4. Measure from one piece of tape to the other that marked your outstretched hands. It was how many centimeters?

5. Determine the area you would cover if you were the shape of the rectangle marked by the tape on the wall.

6. Were you a square?

TEACHING NOTES

Students may express their measurements in two units such as 1 m 20 cm. This does not take advantage of the simplicity of the metric system.

EXTENSION

Determine the area on the wall by using an overhead projector and shining a transparent grid with traced outline. Will the area vary as the distance of the overhead from the wall changes? How would you obtain standard measurements using this procedure?

can you match the area?

TEACHING OBJECTIVE

The students will be able to approximate the area of figures.

PROBLEM

Estimate the areas of figures given and verify your answers by measuring.

NUMBER OF PARTICIPANTS

Any number.

MATERIALS

Prepared worksheet, centimeter grid paper, and carbon paper.

DISCUSSION

If a figure is not a rectangle, how could you accurately measure the area of the figure?

PREREQUISITE SKILLS

None.

DIRECTIONS

1. Match the listed areas with the figures shown in Figure 14.
2. Check your answer by tracing the figures onto the grid paper (Figure 15) using carbon paper and counting the number of square centimeters as accurately as possible.

(a) 25 cm²
(b)   4 cm²
(c) 29 cm²
(d) 13 cm²
(e)   9 cm²
(f) 21 cm²

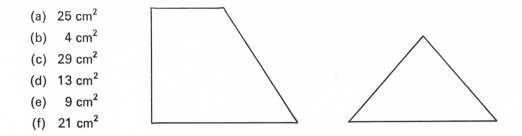

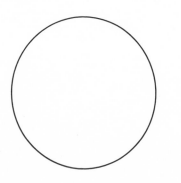

FIGURE 14   Match the area with the figure.

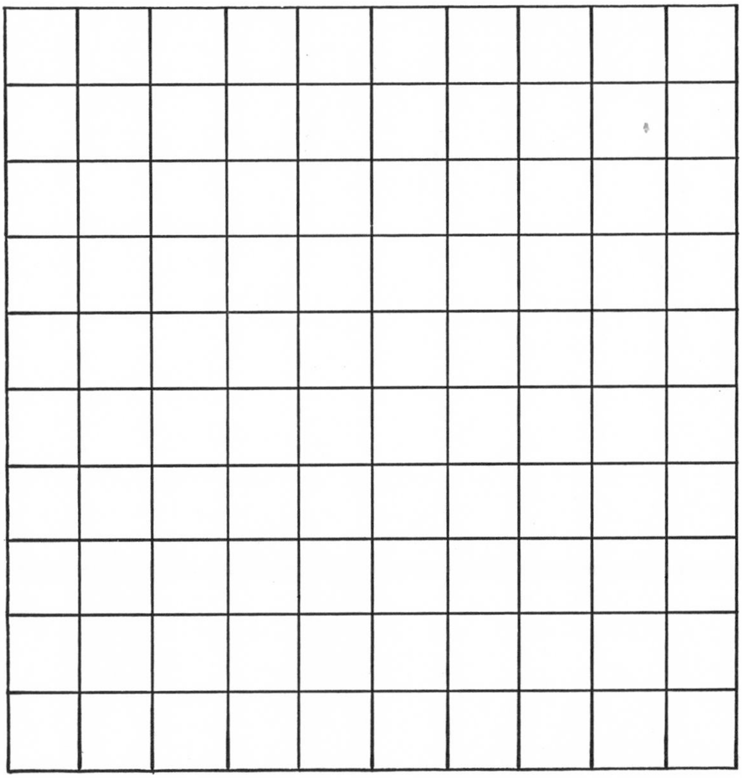

FIGURE 15. Centimeter grid paper.

TEACHING NOTES

Grid paper should be run off in adequate quantities for students to use in further area experiments.

EXTENSIONS

Have students make up triangular designs and estimate their areas by using the centimeter grid paper.

*what is the area and perimeter of
your silhouette?*

TEACHING OBJECTIVE

The student will be able to calculate his or her surface area and perimeter.

PROBLEM

Estimate your surface area and perimeter; then verify your results.

NUMBER OF PARTICIPANTS

Work in groups of two.

MATERIALS

Butcher paper, string, metric ruler, overhead transparency grid, magic marker.

DISCUSSION

How can you find the perimeter and area of an irregular shape?

PREREQUISITE SKILLS

None.

DIRECTIONS

1. Lie down on a piece of butcher paper and have a friend trace your outline with a magic marker.
2. Take a piece of string and lay it along your outline. Measure the length of the string to find the perimeter of your silhouette.
3. Tape your silhouette on a wall and shine the overhead transparency grid on the silhouette, then trace the grid onto the silhouette.

4. Determine the area of each square of your silhouette.
5. Now count the number of squares in your silhouette.
6. Compute the area of your silhouette.

TEACHING NOTES

Student will find the computation much easier if you move the overhead projector until each square on the silhouette is a decimeter square on the silhouette.

EXTENSIONS

Find the average surface area of the members of your class.

could you carpet your school halls?

TEACHING OBJECTIVE

The students will be able to compute the area of various rectangular shapes.

PROBLEM

Determine how many square meters of carpeting you would need to carpet your school's halls.

NUMBER OF PARTICIPANTS

Any number.

MATERIALS

Metric tape measure or trundle wheel.

DISCUSSION

What does it mean to find the area of a rectangle?
Why can you multiply the length by the width of a rectangle to find its area?

PREREQUISITE SKILLS

Ability to use a tape measure or trundle wheel.
Knowledge of formula for finding area.

DIRECTIONS

1. With your trundle wheel or tape measure find the dimensions of the halls in your school.
2. Calculate the area of all of the halls. How many square meters of carpeting would you have to buy to carpet your school's halls?

3. If the school could get the carpeting for $9 per square meter, how much would it cost to carpet your school halls?

EXTENSION

Students could find out how much it would cost to carpet all of the classrooms.

are sports metric?

TEACHING OBJECTIVE

The student will be able to use a trundle wheel or metric tape measure to measure long distances.

PROBLEM

Make baseball, basketball, and football metric sports.

NUMBER OF PARTICIPANTS

Small groups work best.

MATERIALS

Trundle wheel or metric tape measure.

DISCUSSION

Since all international sports are metric, will metrication affect American sports?

PREREQUISITE SKILLS

Ability to use trundle wheel or measuring tape.

DIRECTIONS

1. Using the trundle wheel or metric tape, measure the baselines on a ball diamond.
   a. How far in metric units would you have to run if you hit a home run?
   b. How far in metric units would you have to hit a ball to hit a metric home run at the local ball field?

2. Measure the basketball court.
   a. What is the metric area of the court?
   b. How far in metric units would you have to run if you ran 20 laps around the perimeter during practice.
3. Using the trundle wheel or metric tape, measure your football playing area.
   a. What is the area of the football playing field?
   b. What dimensions would you recommend for a metric football field?

TEACHING NOTES

Have students express dimensions in general units instead of exact metric measurements. For example, a home run will be how many meters in length?

EXTENSION

Help students set up a "metric" track meet for your class or school.

*what is the shape of your school
ground?*

TEACHING OBJECTIVE

The student will be able to measure large areas.

PROBLEM

Make a metric scale drawing of your school grounds and, using this drawing, find the area of the grounds.

NUMBER OF PARTICIPANTS

Groups of three work best.

MATERIALS

Trundle wheel or metric tape measure.

DISCUSSION

If the school grounds were not uniform in shape, what procedures could be used for measuring it?

PREREQUISITE SKILLS

Ability to measure with a trundle wheel.
Facility in determining area and perimeter.
Knowledge of scaling.

DIRECTIONS

1. Find the perimeter of your school ground using the metric trundle wheel.
2. Make a metric scale drawing of your school ground.

3. From your scale drawing find the area.
4. Check your scale drawing with that of other groups to see if they agree.
5. Find the average of the different areas of measurement obtained.
6. Compare the average area to the actual area (if available).

TEACHING NOTES

It would be good to check your metric measurements against information the principal can provide. Be sure to use the conversion charts in Appendix E to change the measurements to metric so the comparison will be valid. Do not have students do this conversion!

EXTENSIONS

Draw a metric scale drawing of your school.

how do you predict?

TEACHING OBJECTIVE

The student will be able to determine the number of small unit areas in a larger area.

PROBLEM

Estimate and then verify the number of specified small unit areas in a given larger area.

NUMBER OF PARTICIPANTS

Any number.

MATERIALS

Centimeter grid paper.

DISCUSSION

How would you go about estimating the number of unit squares in a rectangle? the number of unit rectangles in a rectangle?

PREREQUISITE SKILLS

Understanding of the concept of area.

DIRECTIONS

1. Draw a rectangle 10 cm by 16 cm on the grid paper.
2. Estimate or calculate the number of 3 cm by 4 cm rectangles you can cut from the rectangle.
3. Try it!
4. Were you right?

5. Repeat steps 1, 2, and 3 and this time calculate the number of 2-cm squares you can cut from the rectangle.

TEACHING NOTES

Have each student keep a record of the number of rectangles he or she cuts to see who can get the most.

EXTENSION

Students can work up variations of their own. Have them try for a minimum number and a maximum number.

can you make the metric squares?

TEACHING OBJECTIVE

The student will be able to predict the area of squares by knowing the measurement of one side of the square.

PROBLEM

By knowing the length of one side of a square, determine its area.

NUMBER OF PARTICIPANTS

Any number.

MATERIALS

Grid paper.

DISCUSSION

Why can you express 6×6 as 6^2?

PREREQUISITE SKILLS

Knowledge of the multiplication facts.

DIRECTIONS

1. Draw a square 1 cm in length on each side.
2. Determine its area.
3. Draw squares 2, 3, 4, 5, and 6 cm on each side.
4. Determine the area of each square.
5. Predict the area of squares that measure 8, 10, and 12 cm on each side.

TEACHING NOTES

Point out that for teaching beginning area measurement we use figures we can express as squares, but area measurement need not be restricted to squares.

EXTENSIONS

Have students see if a pattern exists for equilateral triangles with side length of 1 cm.

does the surface area change?

TEACHING OBJECTIVE

The student will be able to accurately determine the surface area of a rectangular solid.

PROBLEM

Determine if the surface area changes as the shape of the figure changes.

NUMBER OF PARTICIPANTS

Any number.

MATERIALS

Twelve centicubes or centimeter rods for each player.

DISCUSSION

When the shape of a figure changes, how does this affect the surface area?

PREREQUISITES

Understanding of the concept of surface area.

DIRECTIONS

1. Place 12 cubes or rods in an arrangement as shown in Figure 16.
2. How many squares 1 cm X 1 cm are on the outside surface of the figure? (Don't forget the bottom.)
3. Change the shape of the figure. Did the surface area remain the same?
4. What would be the largest area that can be exposed by one solid figure made up of 12 centicubes or rods?

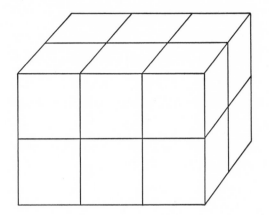

FIGURE 16. 12 cubes.

5. What would be the smallest area that can be exposed by one solid figure made up of 12 centicubes or rods?

6. Repeat with 16 centicubes or rods.

TEACHING NOTES

If neither of these materials are readily available, sugar cubes will work very well.

EXTENSION

Have the students generalize when they will obtain a maximum and minimum surface area.

which gives more—a square or a
rectangle?

TEACHING OBJECTIVE

The students will be able to show that perimeter measurements are independent of area measurements.

PROBLEM

Determine if there is a relationship between the perimeter and area of an object.

NUMBER OF PARTICIPANTS

Any number.

MATERIALS

Centimeter grid paper or metric ruler.

DISCUSSION

How are perimeter and area related?
What shape gives the largest area? Largest perimeter?

DIRECTIONS

1. Draw a series of rectangles on the centimeter grid paper, all having a perimeter of 36 cm, but having the following dimensions:

	perimeter	*area*
a.	1 cm by 17 cm	
b.	2 cm by 16 cm	
c.	3 cm by 15 cm	
d.	4 cm by 14 cm	
e.	5 cm by 13 cm	

f. 6 cm by 12 cm

g. 7 cm by 11 cm

h. 8 cm by 10 cm

i. 9 cm by 9 cm

2. Find the perimeter and area of each rectangle and record it in the chart.

3. Repeat experiment but this time keep the area constant. What happens to the perimeter?

TEACHING NOTES

Have each student make a chart to record the results. Discuss what is the best way to represent the information so a conclusion can be made. Guide the students to make conclusions based on their findings.

EXTENSION

Have students try other shapes to see if the square gives the most area for the least perimeter. Graph the results.

where did the area go?

TEACHING OBJECTIVE

The student will be able to demonstrate that area is independent of perimeter.

PROBLEM

Determine if there is a relationship between the perimeter and area of an object.

NUMBER OF PARTICIPANTS

Any number.

MATERIALS

Twelve soda straws cut to 10-cm lengths for each student.

DISCUSSION

What metric unit is shorter than a meter but longer than a centimeter?

PREREQUISITE SKILLS

Ability to determine area and perimeter.

DIRECTIONS

1. Use 12 of your "decimeter" straws to form a square.
   a. What is its perimeter?
   b. What is its area?
2. Make a figure that has the same perimeter but with an area 1 dm^2 (100 cm^2) less than the original square. What is its area?

3. Make a figure that has the same perimeter but with an area 2 dm², 3 dm², 4 dm², etc., less than the original square.

4. What is the smallest area you can make with a perimeter of 12 dm (120 cm)?

TEACHING NOTES

The use of decimeter-length soda straws is for manipulative convenience only.

EXTENSIONS

Repeat experiment with 20-dm straws.

how long is the path?

TEACHING OBJECTIVE

The student will be able to measure area and linear distance accurately.

PROBLEM

Estimate and then verify the area and length of a given sheet of paper.

NUMBER OF PARTICIPANTS

Any number.

MATERIALS

Adding machine tape or strips cut from butcher paper.

DISCUSSION

What is your prediction of what will happen to the strip as you cut along the lines? Why?

PREREQUISITE SKILLS

Ability to calculate area.

DIRECTIONS

1. Use a piece of adding machine tape or other paper about 70 cm long and 5–7 cm wide.
2. Find the area of the paper.
3. Find the perimeter of the paper.
4. Give the paper a half-twist and tape the ends together (see Figure 17).

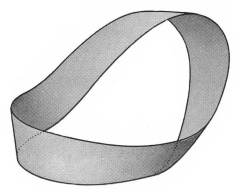

FIGURE 17. Möbius strip.

5. Draw a line down the center of the paper. What happened?
6. Predict what will happen to the strip of paper when you cut on the line.
7. Find the area of the resulting piece or pieces in square centimeters.
8. Find the perimeter

TEACHING NOTES

Advanced students may be interested in doing some research in topology in general and möbius strips in particular.

EXTENSION

Perform the same activity with a full twist in the paper

are 20 questions enough?

TEACHING OBJECTIVES

The students will be able to communicate using metric measure.

PROBLEM

Find the area, perimeter, and shape of a specified geometric shape by asking 20 questions.

NUMBER OF PARTICIPANTS

Two.

MATERIALS

Centimeter grid paper.

DISCUSSION

Name some geometric shapes.
What is the perimeter of a polygon?
How do you determine the area of various polygons?

PREREQUISITE SKILLS

Understanding of the basic knowledge of metric language.

DIRECTIONS

1. One player draws a simple geometric shape (square, rectangle, or triangle) on a piece of the centimeter grid paper.
2. The second player gets 20 questions to try to determine the area, perimeter, and shape of the geometric shape.
3. The first player can only answer yes or no.

4. If the second player does guess correctly within 20 trys, it becomes his or her turn to draw the figure. If not the first player draws a figure again.

TEACHING NOTES

This can be a good activity for accelerated students who need extra challenging activities.

EXTENSION

Have students set up a logical questioning procedure to determine the area, perimeter, and shape of the geometric figure.

can you close the figure?

TEACHING OBJECTIVE

Reinforcing activity in learning to draw line segments to the nearest centimeter, and learning to find area by counting.

PROBLEM

To win the game, get the largest total when adding the area of all the polygons you can construct.

NUMBER OF PARTICIPANTS

Two to four.

MATERIALS

Centimeter dot paper, pencils, metric ruler, and teacher-prepared cards.

DISCUSSION

What is a closed polygon?

PREREQUISITE SKILLS

Ability to read and use a metric ruler.
Ability to follow directions.

DIRECTIONS

1. Shuffle the cards and place them face down.
2. Each player draws a card from the deck to determine the order of play. The player with the highest card goes first.
3. The first player draws a card and constructs a line segment equal to the one stated on the card anywhere on the paper. Line segments

can be constructed either horizontally or vertically but not diagonally.

4. The second player draws a card and constructs a straight line segment, equal to the one stated on the card, anywhere on the paper. Lines may cross but may not overlap each other.

5. Play proceeds until a player is able to construct a closed figure.

6. Once a closed figure is constructed, and the area is found, the player puts the area down on his or her score.

7. Game continues until no further closed figures can be constructed.

8. Winner is the player with the highest score.

TEACHING NOTES

The area is found in this game by counting squares, but this might be a good way to introduce the formula for area.

On cards or 3 cm X 5 cm poster board place the following information, one measurement to a card (number of cards in parentheses):

1 cm (5)	8 cm (2)	50 mm (2)
3 cm (5)	10 cm (2)	6 cm (2)
4 cm (5)	10 mm (5)	30 mm (5)
7 cm (2)	40 mm (5)	9 cm (2)
2 cm (5)		

EXTENSION

Lines can also be constructed on the diagonal to make the game more challenging.

where is the decimal point?

TEACHING OBJECTIVE

The student will be able to correctly change from one metric unit to another for both perimeter and area measurements by moving the decimal point.

PROBLEM

Make each statement true by putting the decimal point in the correct place.

NUMBER OF PARTICIPANTS

Any number.

MATERIALS

Prepared worksheet.

DISCUSSION

How does changing from one metric unit to another differ if you are measuring perimeter versus if you are measuring area?

PREREQUISITE SKILLS

Understanding of multiples of 10.
Knowledge of the relationships between metric units.

DIRECTIONS

Work the following problems by putting the decimal point in correctly to make the statements true. *Be sure to watch the units closely.*

1. $1 \text{ m}^2 = 1\,000\,000 \text{ cm}^2$

2. $1 m^2 = 1\,000\,000\,mm^2$
3. $1 m = 1\,000\,000\,cm$
4. $1 m = 1\,000\,000\,mm$
5. $1\,hectare = 1\,000\,000\,m^2$
6. $1\,mm^2 = 000\,0001\,cm^2$
7. $1\,mm^2 = 000\,0001\,m^2$
8. $1\,cm^2 = 000\,0001\,m^2$
9. $1\,cm = 000\,0001\,m$
10. $1\,mm = 000\,0001\,cm$
11. $1\,mm = 000\,0001\,m$
12. $1\,m^2 = 000\,0001\,hectare$

TEACHING NOTES

Stress the difference between linear and square unit relationships.

EXTENSION

Have the students express the answers in scientific notation.

can you guess?

TEACHING OBJECTIVE

Students will be able to read metric dimensions and communicate them to their fellow students.

PROBLEM

Be the first team to correctly draw a specified geometric shape. Use only metric dimensions to communicate with your team members.

NUMBER OF PARTICIPANTS

Students will work in teams of 12.

MATERIALS

Centimeter grid paper.

DISCUSSION

What makes it difficult to communicate effectively?

PREREQUISITE SKILLS

Understanding of the basic knowledge of metric language.

DIRECTIONS

1. Teacher or one student will draw a simple geometric shape such as a square, rectangle, or triangle on centimeter grid paper.
2. Members of each team will sit back to back.
3. One member of each team will be handed a copy of the grid paper with the geometric shape on it.
4. This player will describe *only* area, perimeter, and location given in metric units, not shape details to the other team members.

5. Without seeing the figure the second team member will try to draw the geometric shape exactly as it is on the first team member's paper.

6. The first team to correctly complete the drawing wins.

TEACHING NOTES

All directions must be in metric measurements and no shape details can be given.

EXTENSION

Discuss the problems students encountered trying to communicate effectively with each other. How could these problems be alleviated?

which triangles fit?

TEACHING OBJECTIVE

The student will be able to make accurate estimates of area sizes.

PROBLEM

Estimate and then verify which two triangles (when fitted together) will form one of the given rectangles.

NUMBER OF PARTICIPANTS

Any number.

MATERIALS

Teacher-prepared worksheet.

DISCUSSION

How many ways can you divide a rectangle into two triangles? Will the triangles be equal?

PREREQUISITE SKILLS

None.

DIRECTIONS

1. Guess which two triangles on the right of Figure 18 will fit together to form the rectangles on the left.
2. Measure the rectangles and triangles shown.
3. Do you want to change your original guesses?
4. Draw lines from the two triangles you think will fit together to make a rectangle.

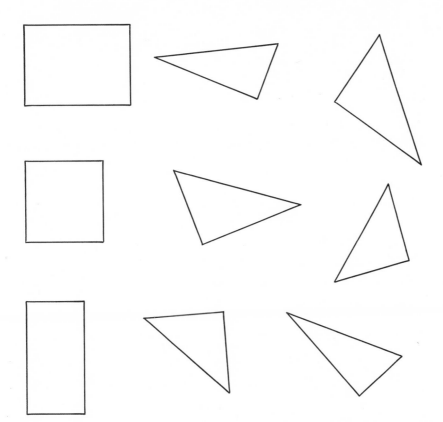

FIGURE 18. Triangle puzzles.

5. Cut out the triangles and see if your estimates were correct.

TEACHING NOTES

Prepare the worksheet on Figure 18 for each student. Discuss ways to estimate which pieces will fit.

EXTENSIONS

Have students do this activity with various polygons.

which is easier?

TEACHING OBJECTIVE

The students will be able to comprehend the advantages of the metric system over the customary system.

PROBLEM

Work the same problem in the customary system and in the metric system. Which was easier?

NUMBER OF PARTICIPANTS

Any number.

MATERIALS

Prepared worksheet.

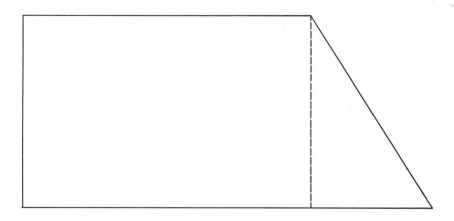

FIGURE 19. Find the perimeter and area.

DISCUSSION

Which fractions do you think will be important when using the metric system?

PREREQUISITE SKILLS

Ability to add fractions.

DIRECTIONS

1. Find the perimeter of Figure 19, first using inches then centimeters.
2. Find the area of Figure 19, first using inches then centimeters.
3. Was it easier to compute the area in inches or centimeters?

TEACHING NOTES

Stress that decimal notation will replace a lot of the operations we have done with fractions in the past.

EXTENSIONS

Convert the perimeter from inches to yards and then from centimeters to meters. Which was the easiest?

fill in the blanks.

1. The measures used in determining surface area will change!
   a. A can of paint will read, "Will cover 20 m² per liter."
   b. If your living room is 4 m × 6 m and carpet is $15 per square meter, what will it cost to carpet your living room? _____
   c. Large areas of land will be measured in square kilometers. How many square meters are there in a square kilometer? _____ (Hint: There are 1000 m in a km.)
   d. Land for homes and farms will be sold in 10 000 m² sections called hectares.
   e. One hectare is equal to approximately 2.5 acres. If you owned 100 acres, approximately how many hectares would you own? _____
   f. Pan sizes will change. A cookie sheet would probably measure _____ .

2. The measures of perimeter will change!
   a. A person sewing will buy binding by the _____ .
   b. A house plan will have the outside dimensions in _____ .
   c. Picture framing will be measured in _____ .

TABLE 6-4.   The gram scale

	*kilogram	(kg)	=	1000	(10^3)	grams
	hectogram	(hg)	=	100	(10^2)	grams
	dekagram	(dag)	=	10	(10^1)	grams
	*GRAM	(g)	=	1	(10^0)	gram
	decigram	(dg)	=	0.1	(10^{-1})	gram
	centigram	(cg)	=	0.01	(10^{-2})	gram
	*milligram	(mg)	=	0.001	(10^{-3})	gram

powers of 10

* Commonly used units

Section III

MASS

Technically, the metric units for measuring mass and weight are different. In common usage, however, the terms mass and weight are often used synonymously and will be so used in this book. It is important to note that the units of measure used in this section are strictly mass units.

Commercial balances and metric mass pieces are utilized in some of the activities. Hopefully, these materials will be available either in the school (check the science laboratory) or school district. If no commercial materials are available the class should make balances as outlined in the activity *"how accurate is a student-made balance?"* The students should also make their own weights (mass pieces) based on the knowledge that a nickel has a mass of 5 g. This is described in the activity *"where can you get mass pieces?"* If students make their own mass pieces be sure to explain that these are not as accurate as commercial mass pieces.

The kilogram is the base unit of mass in the metric system and is usually the easiest to relate to since individual masses will be measured in kilograms. The milligram, a very small unit, is generally used in very technical scientific measurements. In advertisements, the nicotine content of cigarettes is listed in milligrams. Note that the metric ton does not fit the unit prefix format, but does follow the powers-of-10 notation.

To help the beginning student it would be helpful to provide a chart of the mass measures and decimal notations.

TABLE 6-5.   Mass (weight) metric units

Units	Symbol	Multiplication Factor
milligram	mg	0.001 g or 1/1000 g or 10^{-6} kg
gram	g	1000 mg or 1/1000 kg or 0.001 kg
kilogram	kg	10^6 mg or 1 000 000 mg or 1000 g
metric ton	t	10^9 mg or 10^6 g or 1000 kg

exploration

Can you guess the mass?
Which box is the heaviest?
Are small things always light?
What does the graph of the mass of your class look like?
Are the bottles balanced?

discovery

How accurate is a student-made balance?
How many does it take?
Is it more or less?
Where can you get mass pieces?
What mass pieces do you need?
Does popping change the mass of the corn?
Can you use water in place of your mass pieces?
Which dollar's worth of change has the most mass?
Can you explain it?

drill and practice

Does the word make sense?
Can you crack the code?
Can you follow the signs?
Which unit do you name?
Where is the point?
Can you stop in time?

How much are you worth?
What is your mass?
Is this square magic?

general

Applications of mass measures

can you guess the mass?

TEACHING OBJECTIVE

The student will be able to compare the mass of objects and become familiar with the terminology "heavier than, lighter than."

PROBLEM

Compare the mass of given objects.

NUMBER OF PARTICIPANTS

Any number.

MATERIALS

Pan balance (if none is available see activity *"how accurate is a student-made balance?"* to make one), a nickel, various small objects (handkerchief, pencil, chalk, piece of paper, etc.) for each group of three to five students.

DISCUSSION

Which is heavier: a pencil or a piece of chalk?
How can we find out which is heavier?
If we put one object on each side of the balance and one side of the balance goes down, and the other side goes up, which side is heavier?

PREREQUISITE SKILLS

None.

DIRECTIONS

1. Have students compare the mass of the various items.

2. Have students use the nickel as a standard and compare the mass of the various items to the mass of the nickel.

TEACHING NOTES

The students will need instruction on using a pan balance. Many different objects can be used as the standard. Ask the student if this will change the mass of the items in each pile.

EXTENSION

If possible obtain a 5-g mass piece and have students discover that a nickel has a mass of about 5 g. Then use a 5-g mass piece as a standard of comparison to find the mass of various items.

which box is the heaviest?

TEACHING OBJECTIVE

The student will learn that the size and shape of an object does not determine its mass.

PROBLEM

Determine which box is the heaviest and which is the lightest.

NUMBER OF PARTICIPANTS

Any number.

MATERIALS

Pan balance, four boxes same size, one box each for rice, beans, blocks, or puffed rice (one set for every five students).

DISCUSSION

Which box has the greatest mass?
How can you determine which box has the greatest mass?

PREREQUISITE SKILLS

None.

DIRECTIONS

1. Without lifting the four boxes, decide which box is the heaviest, and which is the lightest.
2. Now pick up each box and decide which box is the lightest, and which is the heaviest.
3. Use your balance to determine which box is the lightest, and which is the heaviest.

TEACHING NOTES

The students will need instruction on how to determine which box is heavier or lighter using a pan balance.

EXTENSION

Utilizing only the pan balance, order the boxes from heaviest to lightest.

are small things always light?

TEACHING OBJECTIVE

The student will be able to demonstrate recognition of the noncorrelation of mass and volume by ranking objects according to their mass, not their size.

PROBLEM

Order objects from lightest to heaviest first by sight, then by touch, and then by using a balance.

NUMBER OF PARTICIPANTS

Any number.

MATERIALS

Balance, mass pieces, an assortment of objects such as a lead fish sinker, a nut, a rubber ball, etc.

DISCUSSION

If you had a number of boxes all looking exactly alike, what physical property is most likely to vary?

PREREQUISITES

Understanding of the definition and concept of terms such as larger, smaller, heavier, lighter, etc.

DIRECTIONS

1. Look at the collected objects and in your mind arrange them in a row from the lightest object to the heaviest object without touching them.
2. Rearrange the objects after lifting them with your hands.

3. Rearrange the objects after comparing them on the balance. Use the mass pieces to determine their mass.
4. Did your arrangements change in steps 1, 2, and 3? Why?

TEACHING NOTES

Select a wide range of objects for a more effective activity.

EXTENSION

Have students construct or find objects of different sizes but the same mass and go through the above activities.

*what does the graph of the mass of your
class look like?*

TEACHING OBJECTIVE

The students will be able to determine the mass of each class member and illustrate the tabulation of these measurements on a graph.

PROBLEM

Find the mass of your classmates and then graph the results.

NUMBER OF PARTICIPANTS

Any number.

MATERIALS

Metric kilogram scale and piece of graph paper.

DISCUSSION

What are some of the different ways that information can graphically be presented?

PREREQUISITES

The ability to accurately read a metric kilogram scale and to do basic graphing.

DIRECTIONS

1. Have students determine their mass by using the metric kilogram scale.
2. Then have each student record his or her mass on a graph like the one illustrated in Figure 20.

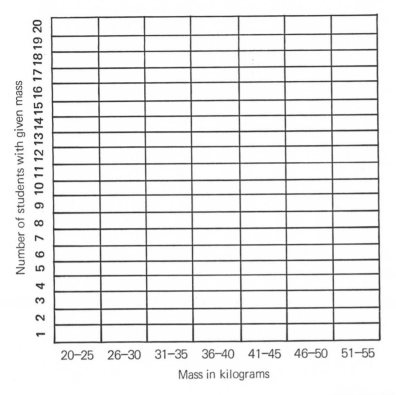

FIGURE 20. Graph the mass of your class.

TEACHING NOTES

If a metric kilogram scale is not available, a customary scale can be used and conversion can be made by dividing the reading in pounds by 2.2.

Make some colored strips of paper or tape cut in rectangles to fit the shape in the chart. If a student weighs 33 kg, a strip of tape or paper would be put in the appropriate column. When every student has recorded his or her mass, a bar graph will be the result.

EXTENSION

This same activity could be used to compare the mass of shoes, books, etc.

are the bottles balanced?

TEACHING OBJECTIVE

The student will demonstrate an understanding of the concept that mass is independent of volume by accurately arranging objects that have the same volume but differing masses in order from lightest to heaviest.

PROBLEM

Order the bottles from lightest to heaviest first by sight, then by touch, and then by using a balance.

NUMBER OF PARTICIPANTS

Any number.

MATERIALS

Balance, mass pieces, and eight to ten baby food jars that have been painted or covered and filled with varying amounts of sand, dirt, or plaster of paris.

DISCUSSION

Are visual observations sufficient to accurately describe an object?
What are some properties of an object that may not be apparent by visual observation?

PREREQUISITES

Ability to use a balance.

DIRECTIONS

1. Try to arrange the bottles from lightest to heaviest by just pointing to them. Do not touch them.

2. Arrange the bottles from lightest to heaviest by picking each bottle up only one time.
3. Arrange the bottles from lightest to heaviest by comparing them as you handle them. Did your arrangements change each time?
4. Arrange the bottles from lightest to heaviest using the balance.

TEACHING NOTES

Discuss why experiences such as these caused man to develop tools to extend his senses.

EXTENSION

Slips of paper can be made, each printed with the correct mass of one of the bottles. The above activities can then be repeated with the students trying to match the bottles with their correct slips.

how accurate is a student-made balance?

TEACHING OBJECTIVE

The students will be able to determine the accuracy of a balance and list factors that affect its accuracy.

PROBLEM

Make your own balance and determine its accuracy.

NUMBER OF PARTICIPANTS

Any number.

MATERIALS

String, small heavy-weight paper plates, a coat hanger, a pencil, and assorted standard mass pieces.

DISCUSSION

What factors influence the accuracy of a balance?
How can you tell when a balance balances?

PREREQUISITES

None.

DIRECTIONS

1. Cut six pieces of string, each piece being 40 cm in length.
2. With a sharp instrument, punch three holes evenly spaced around the paper plates.
3. Knot a piece of the string through each of the holes.
4. Tie the three loose ends of string of one plate to one end of the hanger.

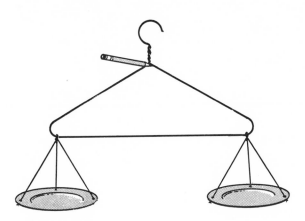

FIGURE 21. A balance.

5. Tie the three loose ends of string of the other plate to the other end of the hanger.
6. Support the hanger on a smooth round object such as a pencil or a nail.
7. Determine if the balance is accurate for a wide range of weights.

TEACHING NOTES

A procedure needs to be worked out so the student will be able to balance the sides.

EXTENSION

Challenge students to construct a device that will accurately weigh objects that have a mass of 0.5 g.

how many does it take?

TEACHING OBJECTIVE

The student will demonstrate the ability to use a balance scale.

PROBLEM

Find different ways of determining the mass of several objects. Graph your results.

NUMBER OF PARTICIPANTS

Any number.

MATERIALS

Balance; centicubes or mass pieces; a number of small objects such as: beans, nickels, pennies, dimes, assorted centimeter rods, paper clips, sheets of writing paper, pieces of chalk, thumbtacks, straight pins, etc.

DISCUSSION

How can one compare the differences in the masses of several objects?

DIRECTIONS

1. Select a group of ten centicubes or a 10-g mass piece.
2. Place the mass pieces or centicubes on one side of the balance and complete the following statements:
    It takes _____ beans to balance ten centicubes.
    It takes _____ straight pins to balance ten centicubes.
    It takes _____ paper clips to balance ten centicubes.
    Etc.
3. Graph the results.

TEACHING NOTES

Let the students select many different objects.

EXTENSION

Have students repeat the experiment using any of the larger mass pieces.

is it more or less?

TEACHING OBJECTIVE

The student will be able to accurately assess whether an object is heavier or lighter than a predetermined mass.

PROBLEM

Decide if an object is heavier or lighter than a predetermined mass and check the accuracy of your decision by using the balance.

NUMBER OF PARTICIPANTS

Any number.

MATERIALS

Balance; centicubes or selected mass pieces; an assortment of small objects such as: a pencil, a quarter, eyeglasses, various centimeter rods, paper clips, a chalkboard eraser, a piece of chalk, etc.

DISCUSSION

How does a person learn to estimate accurately?
Is there a need to have a mental picture of the mass 10 g or 25 g?

PREREQUISITES

None.

DIRECTIONS

1. Select ten centicubes or a 10-g mass piece.
2. Arrange the objects which are lighter than the selected mass piece to the left of it and heavier than the mass piece to the right of it.

3. Check your arrangement on your balance. Reorder them if necessary.
4. Repeat the experiment using a number of different mass pieces.

TEACHING NOTES

Select several groups of objects that the student must work through with the different mass pieces. This will give needed practice in developing a concept of metric mass measurements.

EXTENSION

Try this experiment with the student blindfolded.

where can you get mass pieces?

TEACHING OBJECTIVE

The student will use a pan balance to make reasonably accurate mass pieces.

PROBLEM

Make your own mass pieces and check their accuracy by using a pan balance, water, and a graduated cylinder.

NUMBER OF PARTICIPANTS

Any number.

MATERIALS

Five small plastic vials with cap and sand OR a brick of modeling clay, five nickels, and a reference set of standard mass pieces OR two cups and a graduated cylinder.

DISCUSSION

If the mass pieces you construct are slightly inaccurate, how will this affect your results?

PREREQUISITES

None.

DIRECTIONS

1. Place a vial and its cap on one side of the pan balance.
2. On the other side place a standard metric mass piece that has a larger mass than the vial.

3. Fill the vial carefully with sand, water, or dirt until the vial balances with the standard mass.

4. Assemble the mass piece by putting the cap on the vial.

OR

1. Place a nickel on one side of the balance.

2. Put pieces of modeling clay on the other side until they balance.

3. Mold the modeling clay into a cube and with a pencil label it "5 g".

4. Repeat, using two nickels as the 10-g standard mass piece, four nickels as the 20-g standard mass piece, etc.

5. Check the accuracy of your mass pieces by placing them in a cup on one side of the balance. Pour water into a cup on the other side until the scales balance.

6. With a graduated cylinder measure the amount of water it took to balance the scale. The mass piece and the number of milliliters of water it took to balance them should match.

TEACHING NOTES

Have the students check their mass pieces against both the standard mass pieces and the water if both are available. This will reinforce the relationship that exists between water volume and its mass in the metric system.

EXTENSION

By using uniform strips of cardboard have the students make smaller mass pieces, such as 1-, 2-, or 3-g pieces. Make kilogram mass pieces by using the first procedure and using larger containers.

what mass pieces do you need?

TEACHING OBJECTIVE

The students will be able to determine the minimum number of mass pieces that need to be constructed to determine the mass of any object having a mass from 1 to 100 g.

PROBLEM

Determine the least number of mass pieces you need to weigh any object having a mass greater than 1 g but less than 100 g.

NUMBER OF PARTICIPANTS

Any number.

MATERIALS

None.

DISCUSSION

To find the mass of an object that weighs 6 g do we necessarily need a 6-g mass piece?
What combination of mass pieces could be used instead?

PREREQUISITES

Good facility in simple addition and subtraction.

DIRECTIONS

Given a simple pan balance, determine the least number of mass pieces you need to construct in order to find the mass of any object having a mass greater than 1 g but less than 100 g.

TEACHING NOTES

Careful recording of the mass pieces selected is very important.

EXTENSION

The students may explore what mass pieces are needed to find masses up to a kilogram.

does popping change the mass of the corn?

TEACHING OBJECTIVE

The student will be able to accurately measure mass.

PROBLEM

Find the mass of unpopped corn, then pop the corn and find its mass. Explain the difference between the two masses.

NUMBER OF PARTICIPANTS

Small groups of two to five.

MATERIALS

Balance, mass pieces, popcorn, and popper.

DISCUSSION

What will happen to the volume of the corn during popping?
What do you predict will happen to the mass of the corn? Why?

PREREQUISITES

None.

DIRECTIONS

1. Carefully weigh out approximately 50 g of unpopped corn.
2. Record the mass.
3. Pop the corn.
4. Record the mass of the popped corn.
5. Is there a difference? Why?
6. Eat the corn.

TEACHING NOTES

If the students are quite young, do as a classroom demonstration.

EXTENSION

Have the students perform similar experiments on other foods, such as soaking beans in water.

can you use water in place of your
mass pieces?

TEACHING OBJECTIVES

The student will be able to state the relationship that exists between the mass of a particular amount of water and the volume it occupies.

PROBLEM

Find the relationship that exists between mass and capacity.

NUMBER OF PARTICIPANTS

Any number.

MATERIALS

Pan balance, two paper cups, mass pieces, and a graduated cylinder.

DISCUSSION

Why is it convenient to have a relationship between mass, capacity, and volume in a measuring system?

PREREQUISITES

Ability to use a balance.
Ability to read a graduated cylinder.

DIRECTIONS

1. Place a paper cup on each side of the pan balance.
2. In one paper cup put one of your mass pieces, in the other carefully pour water until the scale is balanced.
3. Measure the amount of water used to balance the scale in the graduated cylinder.

4. Repeat this with several mass pieces.
5. What relationship exists between mass and capacity?
6. Using water, determine the mass of several objects, then check your results using your mass pieces.

TEACHING NOTES

If needed, a baby bottle will serve as a fairly accurate graduated cylinder.

EXTENSIONS

Challenge students to use the reverse of the above procedure in order to make a graduated cylinder from a tall, thin bottle.

which dollar's worth of change has the
most mass?

TEACHING OBJECTIVE

The student will be able to find the mass of objects accurately and calculate using metric measures.

PROBLEM

Find the mass of a dollar's worth of quarters, dimes, nickels, and pennies.

NUMBER OF PARTICIPANTS

Any number.

MATERIALS

Balance, mass pieces, and several quarters, dimes, nickels, and pennies.

DISCUSSION

Is any part of our money system metric?
Would there be advantages to having our coins metric?

PREREQUISITES

Ability to find mass accurately.
Ability to multiply and divide.

DIRECTIONS

1. Find the mass of a dollar's worth of:
    a. quarters
    b. dimes

    c. nickels

    d. pennies

2. If you were to buy coins by mass, which coin would be the best buy?

TEACHING NOTES

Have students find a combination of small coins that would balance larger coins such as the quarter, half dollar, and, if available, the dollar coin.

EXTENSION

Have students find as many combinations of coins as possible that total 50¢. Determine the mass of each combination.

can you explain it?

TEACHING OBJECTIVE

The students will formulate a hypothesis to answer a discrepant question.

PROBLEM

If two cups partially filled with water are evenly balanced, find out if the presence of your finger in one cup of water will cause the balance to go down.

NUMBER OF PARTICIPANTS

Any number.

MATERIALS

Balance, two cups of water.

DISCUSSION

Since your finger is supported by your hand and arm, should it create an imbalance if you put it in a cup of water on one side of a balanced scale?

PREREQUISITES

None.

DIRECTIONS

1. Put two cups partially filled with water on your balance. Make sure they are evenly balanced.
2. Will the presence of your finger in the liquid cause the balance to go down? Your finger is not to touch the cup at any time.

3. Record what you think will happen. Write a justification for your answers.
4. Now try putting your finger in the water.
5. Try again to explain what happened.

TEACHING NOTES

Students may attempt to measure any difference experienced. This could be done by inserting centimeter rods to a predetermined depth and then balancing the scale with mass pieces on the other side.

does the word make sense?

TEACHING OBJECTIVE

The student will be able to recognize the words associated with determining mass.

PROBLEM

Unscramble the letters to form words having something to do with measuring mass.

NUMBER OF PARTICIPANTS

Any number.

MATERIALS

Prepared worksheet.

PREREQUISITES

None.

DIRECTIONS

1. List all of the special words you can think of that are associated with measuring mass.
2. Look at the list of scrambled words and see if you can recognize and unscramble them.
3. Compare your first list and the list of unscrambled words. Who had the most complete list?
   a. acels   *scale*
   b. laacbne   *balance*
   c. sasm   *mass*
   d. gkrialom   *kilogram*

e.	mrag	*gram*
f.	asms psceei	*mass pieces*
g.	lmlairgmi	*milligram*

TEACHING NOTES

Which prefixes are used in connection with other metric words?

EXTENSIONS

Have students scramble the above words again and see if other students can unscramble them.

can you crack the code?

TEACHING OBJECTIVE

The student will be able to perform the basic operations of mass measurement.

PROBLEM

Find the answer to the riddle by solving the given problems.

NUMBER OF PARTICIPANTS

Any number.

MATERIALS

Prepared worksheet.

DISCUSSION

How many ways can a given mass measurement be expressed?

PREREQUISITES

Ability to combine metric units.

DIRECTIONS

1. "Why did the man throw away his electric toothbrush?" To find the answer decode the message.
2. Work the problems and then place the letter on the correct blank.

—— ————— ———— —— ———————— —————
6 1   9 29 11 4   6 3 8 1   3 11   1 7 15 4 10 2 5   4 12 12 4 6

1. 300 g + 0.1 kg = _____ g
2. 300 g + 0.1 kg = _____ kg
3. 1.2 kg + 800 g = _____ g
4. 1.2 kg + 800 g = _____ kg
5. 0.5 t + 500 kg = _____ kg
6. 0.5 t + 500 kg = _____ t
7. 0.25 t + 250 kg = _____ kg
8. 0.25 t + 250 kg = _____ t
9. 2000 g + 1.7 kg = _____ g
10. 2000 g + 1.7 kg = _____ kg
11. 2.5 g + 3.5 g = _____ g
12. 500 mg + 500 mg = _____ g

A = 2000 g
C = 1000 kg
D = 3700 g
E = 400 g
H = 1 t
I = 0.4 kg
L = 500 kg
N = 6 g
O = 1 g
R = 3.7 kg
T = 2 kg
V = 0.5 t

TEACHING NOTES

Make sure that students understand how to add mass units.

EXTENSIONS

Have students make up their own decoding messages.

can you follow the signs?

TEACHING OBJECTIVE

ϰ   The student will be able to correctly compare metric masses using the relationship symbols $<$, $>$, and $=$.

PROBLEM

Insert the proper sign to make the number sentences true.

NUMBER OF PARTICIPANTS

Any number.

MATERIALS

Prepared worksheet.

DISCUSSION

What are the advantages of the metric system over the customary system when changing units within the system?

PREREQUISITES

Knowledge of metric conversion relationships.

DIRECTIONS

1. Express both sides of the number sentence in the same unit.
2. Place one of the signs ($<$, $>$, $=$) in the blank to make the number sentence true.
   a. 0.5 kg _____ 300 g + 140 g + 60 g
   b. 862 g _____ 1 kg – 140 g
   c. 3 kg _____ 2 kg + 800 g

d. 200 g + 100 g + 50 g _____ 200 g + 100 g + 0.08 kg
e. 427 g + 23 g _____ 0.4 kg + 0.05 kg
f. 8 kg + 1100 g _____ 9.2 kg
g. 300 g + 0.1 kg + 1.25 kg _____ 2200 g
h. 1000 g _____ 1 kg
i. 850 kg _____ 0.95 t
j. 400 g + 0.1 kg _____ 0.5 kg

TEACHING NOTES

Construct a variety of worksheets to give needed drill and practice on the basic metric mass relationships.

EXTENSION

Have students solve the above problems using scientific notation and powers of 10.

which unit do you name?

TEACHING OBJECTIVE

The student will be able to change from one metric unit to another.

PROBLEM

Make the conversions correctly to win the game.

NUMBER OF PARTICIPANTS

Two to four.

MATERIALS

Prepared cards.

DISCUSSION

Why is it important to be able to change from one unit to another?

PREREQUISITES

Ability to change metric units and add.

DIRECTIONS

1. Draw one card from each of the three piles.
2. Change the mass measurement named on the first two cards to the unit named on the third card.
3. If the player makes the conversion correctly, that measurement is added to the player's total.
4. The first player to reach a total of 10 metric tons is the winner.

Prepare cards with the following units and numbers. As players begin play put numbers in one pile and divide units into two piles.

Numbers			Units	
5	40	200	milligrams	(5 cards)
10	45	225	grams	(5 cards)
15	50	250	kilograms	(10 cards)
20	100	275	metric ton	(1 card)
25	125	300		
30	150			
35	175			

EXTENSION

Express numbers in powers of 10 and play the above game.

where is the point?

TEACHING OBJECTIVE

The student will be able to correctly place the decimal point so the measurement makes sense.

PROBLEM

Determine where the decimal point should be placed so the measurement is logical.

NUMBER OF PARTICIPANTS

Any number.

MATERIALS

Prepared worksheet.

DISCUSSION

How important is the placement of the decimal point?

PREREQUISITES

Basic understanding of common mass measurements.

DIRECTIONS

1. Place the decimal point in the measurement so that the measurement is logical and is one that would normally be found.
2. In some instances zeros will need to be added.

   a. A sack of grain = 5000 kg
   b. A beef roast = 2000 kg

c. An aspirin tablet = 100 g
d. A head of cabbage = 6 g
e. A truckload of hay = 2000 t
f. A newborn baby = 20 kg
g. A whale = 60 000 t
h. A box of cereal = 45 g
i. An elephant = 4 t
j. A man = 8000 kg
k. A gold nugget = 4000 g
l. A station wagon = 200 t
m. Salt in a cake recipe = 20 kg
n. Meat in a hamburger = 200 kg

TEACHING NOTES

Discuss the units in relationship to grocery store items. Collect grocery store labels to verify some of the units.

EXTENSION

Have students make up similar problems.

can you stop in time?

TEACHING OBJECTIVE

The students will be able to recognize various ways of representing mass and will be able to find the sum of these mass measures.

PROBLEM

Be the winner by accumulating the most points. Be careful so you are not "wiped out."

NUMBER OF PARTICIPANTS

Two to four.

MATERIALS

Game sheet and pencil for each player, specially prepared cubes or cards.

PREREQUISITES

Ability to change from one metric measure to the other.
Ability to perform decimal addition.

DIRECTIONS

1. The first player throws a cube or draws a card and records the result on the game sheet by #1.
2. At this time the player decides whether to quit or to continue playing. If he or she plays and a "WIPE-OUT" results, he or she must cross out all results below his or her last subtotal.
3. When the player elects to quit or a "WIPE-OUT" occurs the next player gets a turn. If he or she elects to quit he or she records the sum of all plays in the subtotal column.

4. The first player getting a total of 1 kg or more wins the game.

TEACHING NOTES

The cube should contain the following on it's faces: 50 g, 0.2 kg, 100 g, 0.15 kg, 25 g, "WIPE-OUT".

If a blank cube is not available, cover a die with masking tape and write on it with a felt pen.

If the game is to be played using cards, a wide variety of cards can be constructed to make the game more challenging. Be sure to have a "WIPE-OUT" card for every five measurement cards.

EXTENSIONS

Have students play game with three "WIPE-OUT" cards for each three metric cards. Does the game strategy change?

**Game sheet for
"Can you stop in time?"**

Game 1	Game 2	Game 3	Game 4
1. _____	1. _____	1. _____	1. _____
Subtotal _____	Subtotal _____	Subtotal _____	Subtotal _____
2. _____	2. _____	2. _____	2. _____
Subtotal _____	Subtotal _____	Subtotal _____	Subtotal _____
3. _____	3. _____	3. _____	3. _____
Subtotal _____	Subtotal _____	Subtotal _____	Subtotal _____
4. _____	4. _____	4. _____	4. _____
Subtotal _____	Subtotal _____	Subtotal _____	Subtotal _____
5. _____	5. _____	5. _____	5. _____
Subtotal _____	Subtotal _____	Subtotal _____	Subtotal _____
6. _____	6. _____	6. _____	6. _____
Subtotal _____	Subtotal _____	Subtotal _____	Subtotal _____
7. _____	7. _____	7. _____	7. _____
Subtotal _____	Subtotal _____	Subtotal _____	Subtotal _____
8. _____	8. _____	8. _____	8. _____
Subtotal _____	Subtotal _____	Subtotal _____	Subtotal _____
9. _____	9. _____	9. _____	9. _____
Subtotal _____	Subtotal _____	Subtotal _____	Subtotal _____
10. _____	10. _____	10. _____	10. _____
Subtotal _____	Subtotal _____	Subtotal _____	Subtotal _____

Game 1 _____

Game 2 _____

Game 3 _____

Game 4 _____

Grand Total _____

how much are you worth?

TEACHING OBJECTIVE

The student will be able to work problems involving the change of units from gram to kilogram and back.

PROBLEM

Determine how much money you would have if you were weighing it instead of counting it.

NUMBER OF PARTICIPANTS

Any number.

MATERIALS

None.

PREREQUISITES

Ability to convert grams to kilograms.
Ability to multiply and divide.
Knowledge of your own mass in kilograms.

DIRECTIONS

1. Given that a silver dollar has a mass of 18 g:
   a. Determine how much money you would have if you had a kilogram of silver dollars.
   b. How much would the mass of 1000 silver dollars be?
   c. If you were placed on one side of the balance and the other side were silver dollars, how many silver dollars would it take to balance with you?
2. Answer the above questions using a nickel as the coin involved.

TEACHING NOTES

Have the students determine which coin they would prefer to be balanced against if they could keep their mass in coins. Remember that a nickel has a mass of 5 g.

EXTENSION

Repeat the above activity using a quarter as the coin.

what is your mass?

TEACHING OBJECTIVE

The student will be able to express a given mass measurement in any metric mass unit.

PROBLEM

Find the mass of five of your friends and yourself; then chart your results.

NUMBER OF PARTICIPANTS

Any number.

MATERIALS

Prepared worksheet, metric kilogram scale or customary scale.

DISCUSSION

What is the most efficient way of recording large and small numbers?

PREREQUISITES

Ability to change metric mass units to other metric mass units.
Ability to express numbers in scientific notation.

DIRECTIONS

1. Determine the mass in kilograms of all the students in your group.
2. Have each student fill in Table 6-6 for six friends.

TABLE 6-6. Sample table

Name	Mass in kg	Mass in g	Mass in t	Mass in mg	Scientific notation for mg	Scientific notation for t
Bill	60	60 000	0.06	60 000 000	6×10^7	6×10^{-2}

TEACHING NOTES

To help students realize the usefulness of scientific notation, work some problems using the values in the chart. If customary scales are used, divide the pounds by 2.2 to convert to kilograms.

EXTENSION

Have students find the average mass of each column.

is this square magic?

TEACHING OBJECTIVE

The student will be able to successfully make metric unit conversions and express them using scientific notation.

PROBLEM

Determine if the square is a magic one by finding the sum of the diagonal squares, the horizontal squares, and the vertical squares.

NUMBER OF PARTICIPANTS

Any number.

MATERIALS

Prepared worksheet.

DISCUSSION

What rules apply to changing units within the metric system?
How do these rules effect scientific notation?

PREREQUISITES

Knowledge of metric unit relationships.
Knowledge of scientific notation.

DIRECTIONS

1. Work the problems in Figure 22 and express the answer in scientific notation.
2. Place the exponent in the circle.
3. Find the sum of the *exponents* in each row, column, and diagonal.

7 kg = ____ mg ○	20 000 g = ____ kg ○	0.5 t = ____ mg ○
46 t = ____ g ○	0.1 t = ____ g ○	4 kg = ____ g ○
0.25 kg = ____ g ○	1 t = ____ mg ○	25 kg = ____ g ○

FIGURE 22. A magic square.

4. If the sums are all the same the square is magic.

TEACHING NOTES

The following rules can help you make up magic squares that can provide lots of fun and practice for your students.
1. The number (exponent) in the center of the square is equal to half the sum of the other numbers in the center column, the center row, or the diagonals.
2. The sum for each column, row, and diagonal is equal to three times the number in the center of the square.
3. The number in each corner is equal to half the sum for the opposite short diagonal.

EXTENSIONS

Have students develop their own magic-square problems using the rules given above.

1. The measure used in determining your mass will change!
   a. A person weighing 220 lbs. would have a mass of 100 kg.
   b. A person weighing 100 lbs. would have a mass of 45.5 kg.
   c. If ground round is $1.00 per pound, how much would you expect to pay for 1 kg of ground round? _____ (*Hint:* 2.2 lb. = 1 kg)

2. The staples you buy will have a different measure!
   a. Instead of buying 25 lbs. of flour, you will buy 10 kg.
   b. Sugar, salt, and many other items will be sold by the gram or kilogram instead of the _____.

3. Pounds as well as long and short tons will change.
   a. Your pickup truck will have a mass in kilograms.
   b. Cargo on a ship would be measured in metric tons. One metric ton is 1000 kg.
   c. Sugar beets will be measured in _____.

4. Advertisements will change. Produce will be advertised in cents per kilogram.
   Complete the sale signs for the following:

   a.
   | Special! |
   | Lettuce |
   | _____¢/kg |

   b.
   | SALE! |
   | Potatoes |
   | _____¢/kg |

   c.
   | Great buy! |
   | Watermelon |
   | _____¢/kg |

5. A commentary for the Miss America Pageant may say, "The heaviest Miss America had a mass of 64.4 _____, and the lightest Miss America had a mass of 47.7 _____."

6. Two men talking may be overheard saying, "I have made great progress with my diet this week; I have lost 3 _____."

7. What do you think the saying "An ounce of prevention is worth a pound of cure" might become?

TABLE 6-7.   The liter scale

↑	kiloliter	(kl)	=	1000	(10^3)	liters
	hectoliter	(hl)	=	100	(10^2)	liters
	dekaliter	(dal)	=	10	(10^1)	liters
powers of 10	LITER	(l)	=	1	(10^0)	liter
	deciliter	(dl)	=	0.1	(10^{-1})	liter
	centiliter	(cl)	=	0.01	(10^{-2})	liter
	milliliter	(ml)	=	0.001	(10^{-3})	liter

Section IV

VOLUME AND CAPACITY

The terms volume and capacity are often used interchangeably. For the purposes of this book a distinction will be made. Volume refers to the amount of space an object occupies while capacity refers to the amount a container will hold. Students should be aware that different units of measurement are used when measuring capacity and volume.

Graduated cylinders, centimeter rods, and centimeter cubes are utilized in some of these activities. Hopefully, these materials will be available either in the school or school district. It is possible to make your own graduated cylinder if one is not available. The activity *"what does it take to make a graduated cylinder?"* helps students make an acceptable one.

"how big is a cubic meter?" should be done by all students to give them an idea of the size of that measure. Keep one of the constructed cubic meters in the classroom for a frame of reference.

To help students become proficient in working with capacity and volume units, exercises similar to *"can you match them?"* and many fill-in-the-blank exercises will prove helpful.

Since the centiliter and kiloliter are used in science, you may wish to expand some of the activities to include these units. Also practice on the decimal change from cubic centimeters to cubic meters and vice versa might be needed. A helpful hint is that the exponent "3" of the units requires a change of three places for each unit change.

TABLE 6-8. Commonly used volume and capacity metric units

Units	Symbol	Multiplication Factors
Volume		
cubic centimeter	cm³	10^{-6} m³
cubic meter	m³	10^6 cm³
Capacity		
milliliter	ml	0.001 l or 1/1000 l
liter	l	1000 ml

exploration

Can you fill it?

How many make it tall?

How big is a cubic meter?

Does the tall, thin container hold more than the short, wide one?

discovery

What does it take to make a graduated cylinder?

How big is a liter?

What is the volume of your fist?

What are the capacities?

How much does this one hold?

What are the metric kitchen measures?

What is the volume of the penny?

What is the volume of the box?

What happens to the volume?

Are mass and volume related?

Do the formulas work?

drill and practice

What are the words?

How many milliliters make a liter?
Can you match them?
How big is your name?

general

Applications of volume measures
Applications of capacity measures

can you fill it?

TEACHING OBJECTIVE

The student will be able to determine the capacity of a given container using standard measuring tools.

PROBLEM

Find out how many milliliters of water it will take to fill the various containers.

NUMBER OF PARTICIPANTS

Any number.

MATERIALS

250-ml beaker, 500-ml beaker, liter container, and various containers.

DISCUSSION

How many beakers of water would it take to fill the liter container? How could we check to see if our answer is correct?

PREREQUISITE SKILLS

None.

DIRECTIONS

1. Have students estimate how many small (250 ml) beakers of water it would take to fill a liter container.
2. Check estimate by filling liter container with small beakers of water.
3. Now estimate and check to see how many 250-ml beakers of water it would take to fill various containers.

TEACHING NOTES

This is a concept that may require a great deal of experimentation before the child becomes comfortable with it.

EXTENSION

Students could use the 500-ml beaker as a measuring device.

how many make it tall?

TEACHING OBJECTIVE

The student will be able to predict the number of units needed to make a square into a cube.

PROBLEM

Determine how many squares are needed to make a cube.

NUMBER OF PARTICIPANTS

Any number.

MATERIALS

Centimeter unit rods or blocks, centimeter multibase cubes or blocks.

DISCUSSION

How would you measure a cube?

PREREQUISITE SKILLS

Recognition of basic shapes and solids.

DIRECTIONS

1. Form a square using centimeter units or cubes.
2. Predict the number of rods it would take to make a cube with a height equal to the dimensions of the square.
3. Determine the number of rods used in all.
4. Repeat for a wide variety of square bases.

TEACHING NOTES

A number of these experiences build a foundation from which you can help the students develop a "formula" for finding the volume of a cube. When someone does find a workable rule, name it after him or her (i.e. "Bob's Rule or Kay's Rule for Finding Volume").

EXTENSION

Build bases that are rectangular and then extend them vertically and try to develop a rule for finding the volume of any rectangular solid.

how big is a cubic meter?

TEACHING OBJECTIVE

The student will be able to describe an accurate estimate of a cubic meter.

PROBLEM

Build a cubic meter using cardboard, metric rule, and masking tape.

NUMBER OF PARTICIPANTS

Two to five.

MATERIALS

Cardboard, metric ruler, masking tape, scissors or knife.

DISCUSSION

How does the word cubic relate to measurement?

PREREQUISITE SKILLS

Ability to use a metric ruler.

DIRECTIONS

1. Decide how many sides a cube has.
2. Cut out or cut and tape together six cardboard squares 1 m on each side.
3. Assemble to form a cube.
4. The volume of this cube is 1 m^3.
5. Using the cubic meter you constructed as a guide, estimate the volume of your classroom.
6. Estimate the volume of the inside of a Volkswagon in cubic meters.

TEACHING NOTES

Discuss the fact that a cubic meter does not have to be a cube to have a volume of 1 m³.

EXTENSION

Have students find several items which have a volume of 1 m³. Estimate the volume of the school in cubic meters.

does the tall, thin container hold more than
the short, wide one?

TEACHING OBJECTIVE

The student will be able to accurately find the capacity of a container.

PROBLEM

Determine which container has the greatest capacity.

NUMBER OF PARTICIPANTS

Any number.

MATERIALS

Piece of rectangular construction paper, scissors, tape (scotch or masking), beans, balance, and two identical paper cups.

DISCUSSION

Does a tall, thin container hold more than a short, wide one?

PREREQUISITE SKILLS

Ability to use a balance to find mass.
Ability to measure in centimeters.

DIRECTIONS

1. Cut the piece of rectangular construction paper in half.
2. Length = _____ cm   width = _____ cm
3. Make a round tube by folding one piece of paper the long way. Tape the edge to make it a cylinder. Now use the other half of the paper and make

another round tube, but roll the piece of paper the other way to make a shorter, fatter tube. Tape the edge to make it a cylinder.

4. Place the long tube in a paper cup on the balance. Fill the tube with beans. What is the mass?

5. Place the shorter tube in a paper cup on the balance. Fill the tube with beans. What is the mass?

6. Compare the masses of the two filled cylinders. Do both tubes hold the same amount or is there a difference? Remember the dimensions of the original pieces of paper were the same.

TEACHING NOTES

Is there a difference? Why?

Discuss other ways of comparing the capacity of cylinders.

Discuss what changes if any would take place if rice were used, sand were used, etc.

If a balance is not available, empty one tube into the other to check if there is a difference.

EXTENSION

Find the volume of the containers.

what does it take to make a
graduated cylinder?

TEACHING OBJECTIVE

The student will be able to construct a graduated cylinder.

PROBLEM

Make a graduated cylinder using items found around the house.

NUMBER OF PARTICIPANTS

Three to five.

MATERIALS

Ten nickels, clay, ketchup bottle or bottle of similar shape, balance, two identical paper cups, graduated cylinder.

DISCUSSION

What shape of bottle would be best for making a graduated cylinder?

PREREQUISITE SKILLS

Ability to use a balance.

DIRECTIONS

1. Place a strip of masking tape from the top of the ketchup bottle, or bottle of similar shape, to the bottom.
2. On one side of the balance place a cup and two nickels. On the other side pour water into a cup until it balances.
3. Pour this water into the bottle.
4. Mark the water line and label it 10 ml.

5. Repeat steps 2, 3, and 4 using four, six, eight, and ten nickels until bottle is labeled to 50 ml.

6. Compare your "graduated cylinder" accuracy to a real graduated cylinder.

7. If the results are different, explain why.

TEACHING NOTES

Point out what errors can be made measuring and also what other factors could cause error.

EXTENSION

Explore making a graduated cylinder, using 1-g units such as small paper clips.

how big is a liter?

TEACHING OBJECTIVE

The student will develop a "frame of reference" for a liter.

PROBLEM

Make a container that will hold a liter and compare the liter to the quart.

NUMBER OF PARTICIPANTS

Three or four.

MATERIALS

16 cm X 28 cm paper, 20 cm X 20 cm cardboard, tape (scotch or masking), quart jar, beans or puffed rice.

DISCUSSION

What things will we measure in liters?

PREREQUISITE SKILLS

None.

DIRECTIONS

1. Form a cylinder by taping the 16-cm edges of paper together. This cylinder will hold approximately 1 liter.
2. Fill the cylinder with puffed rice or beans—be sure to hold the 20 cm X 20 cm cardboard at the bottom of the cylinder to keep the puffed rice or beans from spilling.
3. Pour the puffed rice or beans from the cylinder into the quart jar.

4. Which is larger, the liter or the quart jar?

TEACHING NOTES

Discuss milk carton capacities compared to a liter.

EXTENSION

Determine how many liters it takes to fill a half-gallon jar.
Measure a half-gallon container of milk. Can you make a liter container from it?
Discuss what metric-size containers should be used for bottling milk and soft drinks.

what is the volume of your fist?

TEACHING OBJECTIVE

The student will determine the volume of his or her fist.

PROBLEM

Find the volume of your fist in metric units.

NUMBER OF PARTICIPANTS

Any number.

MATERIALS

Plastic calibrated beaker (1000 ml), water.

DISCUSSION

How do we find the volume of an object?
Is there any other method of determining volume?

PREREQUISITE SKILLS

Ability to read a calibrated beaker.

DIRECTIONS

1. Estimate the volume of your fist.
2. To determine approximate volume of your fist, do the following:
   a. Fill the plastic calibrated beaker with water to the 600-ml mark.
   b. Put your fist into the water and check to see how high the water rises.
   c. The difference between the new recorded volume and 600 ml is the volume of your fist.
3. How close were you to your estimate?

TEACHING NOTES

Be sure student understands that the amount of water displaced by his or her fist is equal to the volume of his or her fist.

EXTENSION

Determine the volume of other objects utilizing the same method. Convert the milliliter measure to cubic centimeters.

what are the capacities?

TEACHING OBJECTIVE

The student will be able to estimate the capacity of various containers.

PROBLEM

Find the capacity of various containers.

NUMBER OF PARTICIPANTS

Three to five.

MATERIALS

Graduated cylinder, large jug, pop bottle (any size), juice glass, mustard bottle, vanilla bottle (any size), other bottles.

DISCUSSION

What is meant by volume?
What is meant by capacity?

PREREQUISITE SKILLS

Ability to read a graduated cylinder.

DIRECTIONS

1. Fill a graduated cylinder with water to the 100-ml mark.
2. Pour this into a large jar. This equals 0.1 liter.
3. Pour ten of these 0.1-liter measures into the large jar. This is equal to a liter.
4. Estimate and then measure the capacity of the following:

Capacity

		Estimate	Measure
a.	pop bottle	_____	_____
b.	juice glass	_____	_____
c.	mustard bottle	_____	_____
d.	vanilla bottle	_____	_____
e.	any container	_____	_____

TEACHING NOTES

Discuss the effect of shapes of the bottles on capacity.

EXTENSION

Have students estimate the capacity of very large containers, such as an automobile gas tank, in liters. Then find brochures which list capacities to compare answers.

how much does this one hold?

TEACHING OBJECTIVE

The student will be able to accurately find the capacity of containers in milliliters.

PROBLEM

Estimate, then check your estimate of, the capacity of various bottles.

NUMBER OF PARTICIPANTS

Any number.

MATERIALS

Graduated cylinder and several different-sized bottles.

DISCUSSION

Have students estimate the capacities of the bottles.

PREREQUISITE SKILLS

Ability to read a graduated cylinder.

DIRECTIONS

1. Estimate the capacities of the bottles and order the bottles by capacity from smallest to largest.
2. Find the capacity of each bottle in milliliters.
3. Re-order the bottles if necessary.

TEACHING NOTES

Discuss "frame of reference" for metric capacity measures. Students should pick a common reference to become familiar with capacity measure.

EXTENSION

Have labels with correct capacities written on them. Show bottles one at a time and have students try to place the correct label on each bottle.

what are the metric kitchen measures?

TEACHING OBJECTIVE

The student will be able to use common kitchen measures metrically.

PROBLEM

Determine the capacity of common kitchen measures in metric units.

NUMBER OF PARTICIPANTS

Any number.

MATERIALS

Graduated cylinder, cup, measuring cup, teaspoon, tablespoon, and water.

DISCUSSION

How many teaspoons make a tablespoon?

PREREQUISITE SKILLS

Ability to read a graduated cylinder.

DIRECTIONS

1. Determine how many teaspoons of water it takes to fill the graduated cylinder to 20 ml.
2. How many milliliters are there in one teaspoon? (To obtain answer divide 20 ml by number of teaspoons.)
3. Repeat step 1 with a tablespoon.
4. How many milliliters are there in one tablespoon?

5. What will be the capacity of a metric cup?

TEACHING NOTES

Use a teaspoon as a frame of reference. Teaspoon measure is often expressed in cubic centimeters.
Why? Discuss reasons for using milliliters.

EXTENSION

Explore other kitchen measures for metric capacities.

what is the volume of a penny?

TEACHING OBJECTIVE

The student will be able to find the volume of objects using the displacement of water procedure.

PROBLEM

Estimate and then find the volume of a penny.

NUMBER OF PARTICIPANTS

Two to five.

MATERIALS

Graduated cylinder, ten pennies, and water.

DISCUSSION

If an object floats, what amount of water is displaced?
If an object sinks, what amount of water is displaced?

PREREQUISITE SKILLS

Ability to use a graduated cylinder.
Ability to convert capacity metric measure to volume metric measure.

DIRECTIONS

1. Estimate the volume of one penny, of ten pennies.
2. Fill the graduated cylinder with 20 ml of water.
3. An object displaces its own volume in water if it sinks.
4. Drop one penny into the graduated cylinder. Now what is the reading of the graduated cylinder?

5. Drop all ten pennies into the graduated cylinder. What is the reading of the graduated cylinder?

6. What is the average volume of each penny?

TEACHING NOTES

Discuss with students how you can get more accurate graduated cylinder readings and the metric capacity and volume measures which are equal.

EXTENSION

Have students repeat the experiment with dimes, nickels, and quarters.

what is the volume of the box?

TEACHING OBJECTIVE

The student will be able to accurately find the volume of a box using both customary and metric units.

PROBLEM

Find the volume of a box in the customary and metric systems. Decide which method is easier.

NUMBER OF PARTICIPANTS

Any number.

MATERIALS

Empty box (shoebox is fine), yardstick, and metric ruler.

DISCUSSION

How do you find volume?
What is the formula for finding volume of a rectangular solid?

PREREQUISITE SKILLS

Ability to compute volume using formula $l \times w \times h$.

DIRECTIONS

1. Measure the dimensions of the box using the yardstick.
2. Find the volume of the box using the formula $l \times w \times h$.
3. Record the volume of the box *in cubic yards.*
4. Now measure the dimensions of the box using the metric ruler.
5. Find the volume of the box using the formula $l \times w \times h$.

6. Record the volume of the box *in cubic meters.*

7. Which system was the easier to use?

TEACHING NOTES

   Discuss which measuring system was the easier to use. The most accurate. The fastest.

   Be careful that the students determine the volume of a cubic yard by using 36 in X 36 in X 36 in.

EXTENSION

   Find the volume in cubic feet.
   Find the volume in cubic centimeters.

what happens to the volume?

TEACHING OBJECTIVE

The student will be able to accurately give volume measure as the dimensions are doubled and tripled.

PROBLEM

Estimate what would happen to the volume of a box if you double and then triple its dimensions. Build the boxes and see if you were correct.

NUMBER OF PARTICIPANTS

One to three.

MATERIALS

Paper, scissors, and tape (scotch or masking), puffed rice or sand.

DISCUSSION

When you double the dimensions of a box, what happens to the volume?

PREREQUISITE SKILLS

Ability to measure and construct boxes.

DIRECTIONS

1. Suppose you had a cube that was 1 cm X 1 cm X 1 cm and you filled the cube with rice or sand. How much rice or sand would the cube hold?
2. Now suppose you made each side 2 cm X 2 cm X 2 cm. How much rice or sand would the new container hold?
3. Next suppose you make each side three times as long as the original cube. How much rice or sand would this container hold?

4. Take some paper, scissors, and tape. Make the containers in steps 1, 2, and 3 and check to see if the volume you predicted was correct. Check your answers by filling containers with water then pouring contents into a graduated cylinder. Remember, 1 ml = 1 cm^3.

TEACHING NOTES

Students have a difficult time giving volume when doubling each side. Naturally they think volume doubles. Be sure to have the student actually do the experiment so he or she will observe the actual increase in volume.

EXTENSION

Have the student double one side and find out if the volume doubles. Have the student double two sides and determine the volume.
Have the student predict what would happen to the volume of a cube if you increase each side ten times. Solve the problem.

are mass and volume related?

TEACHING OBJECTIVE

The student will be able to state a metric relationship between mass and volume.

PROBLEM

Determine if there is a relationship between mass and volume.

NUMBER OF PARTICIPANTS

Three to five.

MATERIALS

Ten centicubes, water, balance, metric mass pieces, and graduated cylinder.

DISCUSSION

How can you increase accuracy when measuring with small units?

PREREQUISITE SKILLS

Ability to use a balance.
Ability to read a graduated cylinder.

DIRECTIONS

1. Find the mass of the centicube using the balance. (To be more accurate, find the mass of ten centicubes then divide by 10.)
2. Drop the centicube into a graduated cylinder with 20 ml of water in it. Make sure it sinks.
3. Now what is the reading of the graduated cylinder?

4. Drop four more centicubes into the graduated cylinder.
5. Now what is the reading of the graduated cylinder?
6. What was the average volume of each centicube?

TEACHING NOTES

Be sure to stress that centicubes are carefully constructed so that 1 cm^3 piece has a mass of 1 g.

EXTENSION

Have students determine if this same activity will work using centimeter rods. Why?

do the formulas work?

TEACHING OBJECTIVE

The student will be able to measure and compute the volume of a rectangular solid and a cylinder.

PROBLEM

Find the volume of a brick and a coffee can.

NUMBER OF PARTICIPANTS

Three or four.

MATERIALS

Depth gauge, vernier calipers, diameter gauge, coffee can (any size), and brick (any size).

DISCUSSION

What is π equal to?
What is the difference between volume and capacity?

PREREQUISITE SKILLS

Knowledge of and ability to use volume formulas for cylinders and rectangular solids.

DIRECTIONS

1. Measure the length, width, and height of the brick using the calipers.
2. Determine the volume of materials used in the construction of the brick using the formula $v = l \times w \times h$. If the brick has a hole in it, be sure to subtract the volume of the hole from the total volume.

3. Measure the inside diameter of the coffee can using the diameter gauge.
4. Measure the height of the coffee can using the depth gauge.
5. Compute the volume of the coffee can using the formula $v = (3.14 \times d^2 \times h)/4$ where d is the inside diameter and h is the depth of the can.

TEACHING NOTES

Try to use a brick such as a cinderbrick so the student will need to subtract the open volume from the total volume.

EXTENSION

Find volumes of different kinds of containers.

what are the words?

TEACHING OBJECTIVE

The student will be able to recognize metric vocabulary.

PROBLEM

Unscramble the letters to form words having something to do with volume and capacity.

NUMBER OF PARTICIPANTS

Any number.

MATERIALS

Copy of the scrambled words for each student.

DISCUSSION

What words do we use when measuring volume and capacity?

PREREQUISITE SKILLS

Ability to identify and spell volume and capacity words.

DIRECTIONS

1. Unscramble the following words. Each word has something to do with volume and capacity measurement.
2. Check to see that all words are spelled correctly.

    a. rellimilti   *milliliter*
    b. miccbueert   *cubic meter*
    c. tecccbuemreinit *cubic centimeter*
    d. cirtem   *metric*

e.	ltrei	*liter*
f.	eemsaru	*measure*
g.	lilliretmi	*milliliter*
h.	gdauardet lidcneyr	*graduated cylinder*

TEACHING NOTES

Discuss SI rules on metric vocabulary and symbols.

EXTENSION

Have a contest to see who can unscramble the words in the least amount of time.

Have students develop their own scrambled word lists.

how many milliliters make a liter?

TEACHING OBJECTIVE

The student will be able to add metric units to equal one liter.

PROBLEM

Find the sum of capacity measures closest to 1 liter.

NUMBER OF PARTICIPANTS

Any number.

MATERIALS

Prepared game cards, paper.

DISCUSSION

How do you add milliliters to milliliters?
How many milliliters are there in a liter?

PREREQUISITE SKILLS

Ability to add milliliters to milliliters.
Ability to change milliliters to liters.

DIRECTIONS

1. Prepare for the game by cutting ten pieces of poster board 5 cm X 10 cm. Put the numerals 0 to 9 on the cards (one number to a card).

2. Each player prepares the following grid on his or her other paper:

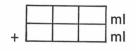

3. The ten cards are shuffled and placed face down on the table.

4. One card is turned over and each player decides where he or she wants to write that number on his or her grid. (He or she is trying to have his or her sum equal one liter.)

5. The next card is turned over and the players now place this number on the grid. The play continues until six numbers have been turned over and each player has placed the numbers on his or her grid.

6. The players now add the milliliter quantities and the player or players closest to one liter receive one point.

7. The cards are reshuffled, new grids are made, and the game proceeds as before. The winner is the first player to get ten points.

TEACHING NOTES

Discuss how to express milliliters in liters and vice versa.

EXTENSION

Play the game so that the student closest to a total of 500 ml receives a point.

Play the game and subtract quantities with the winner being the student with the difference nearest to zero. (Subtract the smallest number from the largest number.)

can you match them?

TEACHING OBJECTIVE

The student will be able to match equivalent volume and capacity measures.

PROBLEM

Match equivalent measures to form a square.

NUMBER OF PARTICIPANTS

One or two.

MATERIALS

Prepared worksheet and scissors.

DISCUSSION

Do you know how to match puzzle pieces?

PREREQUISITE SKILLS

Knowledge of metric volume and capacity measures.

DIRECTIONS

1. Cut Figure 23 into nine pieces.
2. Mix them up.
3. Form the square by matching equivalent volume and capacity measures.

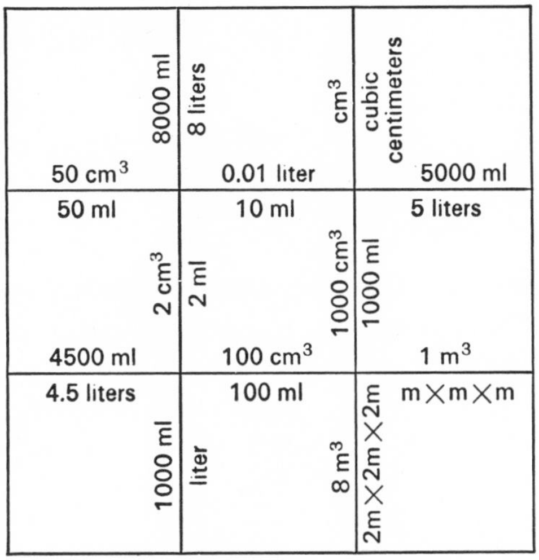

8000 ml · 8 liters · cm³ · cubic centimeters

50 cm³   0.01 liter   5000 ml

50 ml   10 ml   5 liters

2 cm³ · 2 ml · 1000 cm³ · 1000 ml

4500 ml   100 cm³   1 m³

4.5 liters   100 ml   m × m × m

1000 ml · liter · 8 m³ · 2m × 2m × 2m

FIGURE 23. Match the puzzle pieces.

TEACHING NOTES

Paste a picture on the back before cutting Figure 23 into pieces. Student can then turn over the completed puzzle to check the answer.

EXTENSION

Have students make up their own puzzles.

how big is your name?

TEACHING OBJECTIVE

The student will be able to add capacity units, given assigned measurements

PROBLEM

Find the capacity of your name.

NUMBER OF PARTICIPANTS

Any number.

MATERIALS

None.

DISCUSSION

What rules are followed when adding different metric units, e.g. liters, milliliters, etc.?

PREREQUISITE SKILLS

Ability to add capacity and volume metric units.

DIRECTIONS

1. Assign each letter of the alphabet a capacity value.
2. Divide class into two groups.
3. Taking initial of first, middle, and last name of each student, have each student compute their capacity.
4. Have each group total their capacity.
5. Winner is the team with largest total.

TEACHING NOTES

Distinguish between metric capacity and volume units and give their relationship to each other.

EXTENSION

Do activity in volume units only.
Do activity in both volume and capacity units.

fill in the blanks

1. Recipes will change!
   Ingredients will be measured in cubic centimeters (cm³) or in milliliters (ml).
   1 tablespoon salt = _____ cm³ salt
   1 cup flour = _____ cm³ flour
   1 teaspoon baking powder = _____ cm³ baking powder

2. Concrete and soil measures will change!
   Instead of cubic yards of concrete or soil, we will use cubic _____.

3. Cooking utensil sizes will change!
   A 10″ X 6″ X 1½″ cake pan will be 25 cm X 15 cm X 4 cm.
   A 13″ X 9″ X 2″ cake pan will be 33 cm X 23 cm X 5 cm.
   An 8″ X 1½″ round cake pan will be 20 cm X 4 cm.
   An 8″ X 8″ X 2″ square cake pan will be _____ cm X _____ cm X _____ cm.

4. Can you think of other volume changes that will be made?

fill in the blanks

1. Recipes will change:
   Liquids will be measured in liters and milliliters.
   ½ teaspoon vanilla = _____ ml vanilla
   1 qt. boiling water = _____ boiling water
   1 cup shortening = _____ shortening

2. The way you buy your gas and oil will change!
   a. Instead of buying 10 gallons of gas, you would buy 42 liters.
   b. Instead of buying quarts of oil, you would buy _____ of oil.
   c. If you were to go into a gasoline station to buy gas what would you most likely say? (*Hint:* Fill 'er up!)

3. The units to measure milk, juice, and even liquor that you buy will change!
   a. Instead of a pint of cream you would buy _____ of cream.
   b. Instead of half a gallon of milk, you would buy _____ of milk.

4. Measures for small amounts of liquids like perfume will be changed!
   Instead of buying perfume by the ounce, we will buy it by the _____.

Section V

TEMPERATURE

The name "Celsius" has been chosen as the SI metric thermometer since some countries use the term "centigrade" for angle measurements.

It would be preferable to use a Celsius thermometer for the activities in this section. If none is available, use masking tape and convert a Fahrenheit thermometer, using the freezing and boiling points of water as guides. Subdivide the distance between these two points into 100 parts.

The activities *"what is a typical temperature?"* and *"how hot is it?"* should be done by all students to give them some familiarity with common temperatures using the Celsius scale.

To help students relate to Celsius measure, a daily Celsius temperature chart for the school year should be kept.

TABLE 6-9. Temperature Metric Units	
Units	*Symbol*
Celsius	°C

exploration

What makes it hot?

discovery

How hot is it?
How low will the mercury go?
At what temperature does water boil?
What happens as ice melts?

drill and practice

What is a typical temperature?
What is the message?
Can your temperature be predicted?
What do the temperatures say?

general

Applications of temperature measure

what makes it hot?

TEACHING OBJECTIVE

The student will be able to express an understanding that the body is an inaccurate temperature gauge.

PROBLEM

Which hand tells what the temperature really is?

NUMBER OF PARTICIPANTS

One to three at a time, but all should participate.

MATERIALS

Three pans of water, one cold, one lukewarm, and one as hot as the hand can stand.

DISCUSSION

Can you really tell when it is hot or cold? Does being hot or cold make temperature change more noticeable?

PREREQUISITE SKILLS

None.

DIRECTIONS

1. Stick one hand in the cold water and the other hand in the hot water for about three minutes.
2. Then put both hands in the lukewarm water.
3. Describe what you feel.

TEACHING NOTES

Discuss the body's defense and adaptability mechanisms and the need for measuring tools which are not influenced by other factors.

EXTENSIONS

Change the temperature of the lukewarm water and see how your hands react.

how hot is it?

TEACHING OBJECTIVE

The student will be able to name the Celsius temperature reading for four common temperatures.

PROBLEM

Find the Celsius temperature for the boiling point and freezing point of water, a comfortable room, and your normal body temperature.

NUMBER OF PARTICIPANTS

Any number.

MATERIALS

Celsius thermometer.

DISCUSSION

What are the Fahrenheit temperatures for:

1. Boiling water
2. Freezing water
3. Room temperature
4. Body temperature

PREREQUISITE SKILLS

Knowledge of common Fahrenheit temperatures.
Ability to read thermometer.

DIRECTIONS

Using the Celsius thermometer, determine the Celsius temperature for:

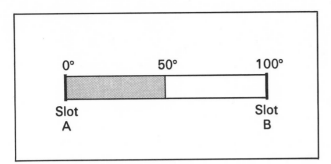

FIGURE 24. "Teaching" thermometer.

1. Boiling water
2. Freezing water
3. A comfortable room temperature
4. Normal body temperature

TEACHING NOTES

It is best if students actually perform experiments to determine the temperatures. However, if this is not possible, have students read a dual-scale thermometer to give them needed practice.

EXTENSIONS

Have students build a thermometer to practice reading temperatures. They could make one by threading a continuous ribbon through poster board and coloring half of the ribbon red. (See Figure 24.)

how low will the mercury go?

TEACHING OBJECTIVE

The student will be able to name one factor that will lower the freezing point of water.

PROBLEM

Find one factor that will lower the freezing point of water.

NUMBER OF PARTICIPANTS

Any number.

MATERIALS

Pan of ice, Celsius thermometer, and a cup of salt.

DISCUSSION

How is ice cream made in an ice cream mixer?

PREREQUISITE SKILLS

Ability to read a Celsius thermometer.

DIRECTIONS

1. Place thermometer in pan of ice water and after several minutes record temperature.
2. Wait ten minutes and record the temperature again.
3. Carefully stir a cup of salt into the pan of ice water.
4. Keep stirring and record the Celsius temperature every minute until there is no change for three minutes.
5. Graph your results.

6. What are your findings?

TEACHING NOTES

Be sure to have an adequate supply of ice. Keep adding ice and salt to the pan until the temperature makes no change for three minutes. The graph can be done as a class activity.

EXTENSION

Look up the history of the Fahrenheit experiment that led to the development of the thermometer. Determine where the name Celsius came from.

at what temperature does water boil?

TEACHING OBJECTIVE

The student will be able to list two factors that influence the boiling point of water.

PROBLEM

Find two factors that influence the boiling point of water.

NUMBER OF PARTICIPANTS

Any number.

MATERIALS

Pan for boiling water, hot plate, Celsius thermometer, sugar (1 cup), and salt (1 cup).

DISCUSSION

At what metric temperature does water boil?

Are there factors that could change the boiling point of water? If so, what are they?

PREREQUISITE SKILLS

Knowledge of laboratory procedure.

DIRECTIONS

1. Put water in the pan and heat the water until it boils. Record the temperature using the Celsius thermometer. Be careful to keep the thermometer off the bottom of the pan.
2. Add salt slowly to the boiling water. Stir constantly and record temperature every minute or two.

3. Record your findings.
4. Repeat steps 1 to 3 using sugar in place of salt.
5. What are your findings?

TEACHING NOTES

This may be an activity that you will want to perform. If you do the activity, have students read the thermometer and record the findings. Discuss the results.

EXTENSION

Determine the length of time it takes to get equal amounts of water from 50°C to 90°C, once with the pan uncovered and once with a lid on the pan.

what happens as ice melts?

TEACHING OBJECTIVE

The student will be able to accurately read a Celsius thermometer.

PROBLEM

At specified time intervals take a reading of a Celsius thermometer that has been placed in a pan of ice and water. Graph your results.

NUMBER OF PARTICIPANTS

Any number.

MATERIALS

Celsius thermometer, pan of ice.

DISCUSSION

As ice melts, what do you predict will happen to the temperature? Draw a predicted time-temperature curve.

PREREQUISITE SKILLS

Ability to read a Celsius thermometer.
Ability to graph information.

DIRECTIONS

1. Place the Celsius thermometer in a pan of ice and water and take a reading.
2. Take a reading every two minutes. Do this five times and record the readings.
3. Now, read the thermometer every 20 minutes.

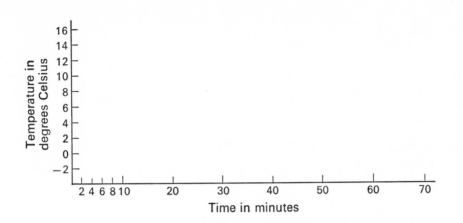

FIGURE 25. Graph of temperature of ice in water.

4. Record the temperatures on the graph in Figure 25. All temperatures should be recorded in degrees Celsius.

TEACHING NOTES

To speed up the process, place the pan in a warm place.

EXTENSION

Boil water and record time-temperature as water reaches its boiling point.

what is a typical temperature?

TEACHING OBJECTIVE

The student will be able to relate Celsius temperature to everyday situations.

PROBLEM

Estimate in Celsius degrees what the typical July temperatures would be in cities marked on the map in Figure 26. Verify your results using an encyclopedia.

NUMBER OF PARTICIPANTS

Any number.

FIGURE 26. What are the typical July Celsius temperatures?

MATERIALS

Prepared worksheet.

DISCUSSION

Will there be more below-zero temperature days in Celsius measurement than there have been in Fahrenheit measurement or vice versa? Why?

PREREQUISITE SKILLS

None.

DIRECTIONS

1. Place what you think would be a typical July Celsius temperature next to the cities marked on the map.
2. Check your answers by looking for typical temperatures in the encyclopedia.

TEACHING NOTES

Be sure to clarify normal climate conditions for any region which the students are unfamiliar with.

EXTENSION

Repeat activity using a world map.

what is the message?

TEACHING OBJECTIVE

The student will be able to compute using positive and negative temperatures.

PROBLEM

Read the message by solving problems and using your results to place the correct letter in each blank.

NUMBER OF PARTICIPANTS

Any number.

MATERIALS

Prepared worksheet.

DISCUSSION

When it goes from −10°C to 15°C, how much has the temperature changed?

When it goes from 20°C to 10°C, what is the change?

PREREQUISITE SKILLS

Ability to compute with integers

DIRECTIONS

1. Solve the problems to get the corresponding letters.
2. Write in the blank the letter that has an answer that matches the temperature under the blank.

__ __ __ __ __ __ __ __ __ __
25 −5 20 15   −10 10   12 15 50 80

__ __ __ __ __ __ __ __ __ __ __ __
−10 −15   5 −15 −12 −20 −5   5 10 −25 20 −5

The temperature changed from:

−10°C to −5°C for A
−12°C to 0°C for C
−40°C to 40°C for D
20°C to 15°C for E
5°C to −5°C for I
−20°C to 30°C for L
0°C to −20°C for M
5°C to −10°C for N
−10°C to 5°C for O
−40°C to −20°C for R
−3°C to 7°C for S
0°C to −25°C for U
12°C to 0°C for Y
−20°C to 5°C for Z

TEACHING NOTES

The rules of signed numbers may have to be reviewed.

EXTENSION

Have students make up similar puzzles for each other.

can your temperature be predicted?

TEACHING OBJECTIVE

The student will be able to use Celsius temperatures.

PROBLEM

Get the correct answer by following directions carefully and doing computations accurately.

NUMBER OF PARTICIPANTS

Any number.

MATERIALS

None.

DISCUSSION

What are some common Celsius temperatures?

PREREQUISITE SKILLS

Ability to do basic computations using integers and squaring numbers.

DIRECTIONS

1. Read the following directions to the students.
   a. Select your favorite Celsius temperature.
   b. Subtract four degrees.
   c. Square the result.
   d. Subtract the square of the original temperature.
   e. Add eight times the original temperature.
   f. What temperature did you get?

g. If you follow the above directions will you always get 16 degrees for your answer?

2. Read the following directions to the students.
   a. Select a Celsius temperature.
   b. Double it.
   c. Increase your result by 40 degrees.
   d. Divide your temperature by 2.
   e. Subtract three degrees more than your original temperature.
   f. Double this result.
   g. Add three degrees to your present answer.
   h. What temperature did you get?
   i. If you follow the above directions will you always get the normal body temperature?

TEACHING NOTES

This procedure can be used for drill and practice in many areas.

EXTENSION

Have students make up similar exercises.
Have students determine why the above exercises work as they do.

what do the temperatures say?

TEACHING OBJECTIVE

The student will be able to perform addition and subtraction operations using Celsius temperatures.

PROBLEM

Find the secret message by solving problems and using your results to place the correct letter in each blank.

NUMBER OF PARTICIPANTS

Any number.

MATERIALS

Prepared worksheet.

DISCUSSION

What rules do you follow when you add and subtract degrees Celsius?

PREREQUISITE SKILLS

Ability to compute with the basic operations.

DIRECTIONS

1. Solve the following problems.
2. Match your answer with the corresponding letter.
3. Put that letter on the appropriate space of the puzzle.
4. When you are finished, a secret message will appear.

    1. $26°C - 3°C =$ $1 = A$

2. $10°C - 9°C =$ 2 = B
3. $10°C + 9°C + 1°C =$ 3 = C
4. $30°C + 15°C - 40°C =$ 4 = D
5. $2°C + 13°C + 3°C =$ 5 = E
6. $(9°C - 5°C) + 2°C =$ 6 = F
7. $(24°C + 12°C) ÷ 2 =$ 7 = G
8. $25°C ÷ 5 =$ 8 = H
9. $(30°C ÷ 2) - 10°C =$ 9 = I
10. $13°C + 13°C =$ 10 = J
11. $19°C - 14°C =$ 11 = K
12. $(2 × 9°C) + 1°C =$ 12 = L
13. $1°C × 1 =$ 13 = M
14. $15°C + 4°C + 1°C =$ 14 = N
15. $3°C + 13°C + 10°C =$ 15 = O
16. $25°C - (10°C + 10°C) =$ 16 = P
17. $(10°C + 5°C + 5°C) - 2°C =$ 17 = Q
18. $3 × 5°C =$ 18 = R
19. $5°C - 1°C =$ 19 = S
20. $10°C - 5°C =$ 20 = T
21. $(2 × 5°C) - 3°C =$ 21 = U
22. $5°C + 13°C =$ 22 = V
23. $3°C + 2°C =$ 23 = W
24. $15°C - 10°C =$ 24 = X
25. $4°C + 15°C =$ 25 = Y
26. $2°C + 1°C =$ 26 = Z

___ ___ ___ ___ ___   ___ ___ ___ ___ ___ ___ ___   ___ ___
 1   2   3   4   5   6   7   8   9 10 11 12   13 14

___ ___ ___ ___   ___ ___ ___ ___ ___ ___ ___   ___
15 16 17 18   19 20 21 22 23 24 25   26

TEACHING NOTES

Remind students of the rules when working with operations and parentheses.

EXTENSIONS

Have students make up their own secret messages and problems.

Temperature measures will change!

1. When it is zero degrees Celsius that means it is going to freeze.
2. Forty degrees below zero will remain the same. *Note:* −40°C = −40°F.
3. If you take your temperature and it is up to _____, call the doctor!
4. When baking in an oven, the Celsius temperature will read approximately one-half the Fahrenheit scale.
5. A high of _____ on the 4th of July in Miami, Florida will take some getting used to.
6. The number of below zero days that occur will _____.
7. Room temperatures will feel good when the thermostat is set at 22°C or _____.
8. When your automobile radiator is boiling, what Celsius temperature will the gauge say?

TABLE 6-10. The metric scale

kilo	meter (km) liter (kl) gram (kg)	=	1000	(10^3)	meters liters grams
hecto	meter (hm) liter (hl) gram (hg)	=	100	(10^2)	meters liters grams
deka	meter (dam) liter (dal) gram (dag)	=	10	(10^1)	meters liters grams
METER LITER GRAM	(m) (l) (g)	=	1	(10^0)	meter liter gram
deci	meter (dm) liter (dl) gram (dg)	=	0.1	(10^{-1})	meter liter gram
centi	meter (cm) liter (cl) gram (cg)	=	0.01	(10^{-2})	meter liter gram
milli	meter (mm) liter (ml) gram (mg)	=	0.001	(10^{-3})	meter liter gram

powers of 10

Section VI

CULMINATING ACTIVITIES

This section includes activities designed to help the students practice skills they have learned in all the other sections. Games and some fun worksheets are provided. This is the first section where all the units are used interchangeably. Remind the students of the relationship between mass, volume, and capacity. All the other prefixes not previously used could be added at this time for a better understanding of the metric system and to illustrate the convenience of systematically changing the names of the units from one to the other, e.g. cm to km, or ml to cl, etc. Some of these units are included in the activity *"do you have it?"* and *"can you find the pattern?"*

"what will the weather be?" requires conversion from the customary to the metric system. Do the conversions for the students or provide them with a table from which to read the appropriate metric values.

A games interest center could be developed utilizing the material in this section. It is also possible for the teacher to convert most common games to metric once all the metric units are known.

It would be helpful to prepare a chart of all the metric units and decimal equivalents for students to refer to as a reference while doing the following activities.

TABLE 6-11. Common Metric units

Linear	Mass	Volume	Capacity	Area	Temperature
millimeter (mm)	milligram (mg)	cubic centimeter (cm³)	milliliter (ml)	square millimeter (mm²)	degrees
centimeter (cm)	gram (g)		liter (l)	square centimeter (cm²)	Celsius (°C)
decimeter (dm)	kilogram (kg)	cubic meter (m³)		square meter (m²)	
meter (m)	metric ton (t)			hectare (ha)	
kilometer (km)				square kilometer (km²)	

drill and practice

What change can a prefix make?
What am I?
Can you match them?
How is your metric vocabulary?
Do you know your metric self?
Can you fill in the blank?
Do you have it?
What is the correct sign?
What would it be like living metrically?
Where is the decimal? What is the word?
How much does it cost?
Can you find the pattern?
What will the weather be?

what change can a prefix make?

TEACHING OBJECTIVE

The student will be able to predict the number of decimal places that will be involved when prefixes are changed.

PROBLEM

Construct a Metric Slide Chart* and use it as a checking device to validate the change of the decimal as prefixes are changed.

NUMBER OF PARTICIPANTS

Any number.

MATERIALS

Prepared worksheet.

DISCUSSION

When changing from one metric measuring unit to another, what is the most important thing to remember?

PREREQUISITES

Ability to work with decimal notations.
Understanding of relationships between the various metric measures.

DIRECTIONS

1. Assemble the Metric Slide Chart as shown in Figure 27.
2. Work the following problems and then use the Metric Slide Chart to check your answers.

   Enter the correct multiplication factor for the unit change:
   a. Meters to centimeters, multiply by _____.

* © 1974 J.C. Penney Company, Inc., Educational Relations Department.

Inner piece

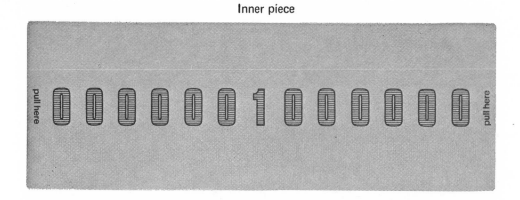

Outer piece

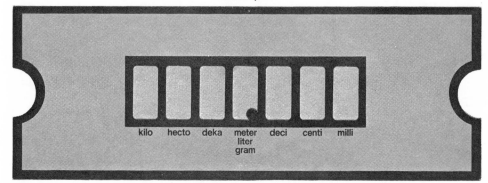

Combined

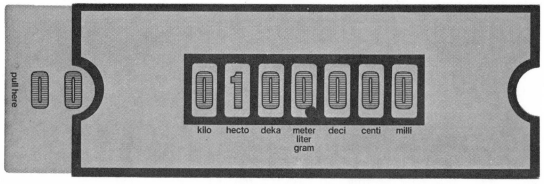

1 hecto(meter) = 100 meters
     (liter)   liters
     (gram)   grams

FIGURE 27. Metric Slide Chart

b. Kilograms to grams, multiply by _____.
c. Liters to milliliters, multiply by _____.
d. Hectometers to centimeters, multiply by _____.
e. Kilograms to milligrams, multiply by _____.
f. Decimeters to millimeters, multiply by _____.
g. Kilometers to centimeters, multiply by _____.

Enter the correct division factor for the unit change:
a. Millimeters to meters, divide by _____.
b. Centimeters to meters, divide by _____.
c. Grams to kilograms, divide by _____.
d. Centimeters to hectometers, divide by _____ .
e. Milliliters to liters, divide by _____.
f. Milligrams to kilograms, divide by _____.
g. Centimeters to kilometers, divide by _____.

TEACHING NOTES

It might be best to enlarge the illustration on a ditto master and then run copies off on oaktag.

To operate the Metric Slide Chart, slide the number 1 to the original unit, then read the multiplication or division factor by covering up all of the zeros past the unit you are converting to.

EXTENSION

Make up problems for the students such as: 4.5 kilograms is equal to _____ grams.

what am I?

TEACHING OBJECTIVE

The student will be able to identify objects *in the room* from a description of their metric measurements.

PROBLEM

Identify an object from a description of its metric measurements.

NUMBER OF PARTICIPANTS

Any number.

DISCUSSION

How do you play the game 20 questions?

PREREQUISITE SKILLS

Basic knowledge of approximate metric measurements.

DIRECTIONS

1. The leader (teacher) describes some object in the room using only its metric measurements. (For example: Its mass is 5 g. It has a length of 3 cm.)
2. The students can ask 15 questions which can be answered only yes or no in order to discover the object.
3. The first person to correctly identify the object gets to describe a new object for the rest of the class to identify.

TEACHING NOTES

Activities where students have had opportunity to learn the measurements of many of the objects in the classroom need to be done prior to this.

EXTENSION

An object could be put in a bag and each student could record its approximate metric measurements by feel. A point is scored for each measurement in which the student has the most accurate measurement. The student with the most points is the winner.

can you match them?

TEACHING OBJECTIVES

The student will reinforce his or her knowledge of the correct symbol with the correct metric word.

PROBLEM

Collect the most cards by remembering where the metric word card and its symbol card are located.

NUMBER OF PARTICIPANTS

Two or three.

MATERIALS

Teacher-prepared cards.

DISCUSSION

Why are symbols used so often in mathematics?
How much care must be used to put symbols with numbers?

PREREQUISITE SKILLS

Knowledge of metric terms and their symbols.

DIRECTIONS

1. Shuffle cards and place them face down (four rows, six cards in a row).
2. The first player turns over two cards, one at a time, trying to match a metric word and its symbol.
3. If the player is successful, he or she keeps both cards and takes another turn.

4. If the player is unsuccessful, the two cards are returned to their original playing position face down.
5. Each player proceeds as the first player did following steps 2, 3, and 4.
6. Play continues until playing board is cleared. The winner is the player with the most cards.

TEACHING NOTES

Prepare the following cards, one unit to a card (*Note:* If blank cards are not available, cut 24 pieces of posterboard, each card 5 cm by 10 cm):

centimeter	cm	gram	g
meter	m	kilogram	kg
millimeter	mm	decimeter	dm
kilometer	km	metric ton	t
degrees Celsius	°C	cubic centimeter	cm³
liter	l	milliliter	ml

If necessary a chart could be provided listing metric words and the appropriate symbols.

EXTENSIONS

Add more words and their symbols to the game (e.g., square millimeter, square centimeter, etc.).

how is your metric vocabulary?

TEACHING OBJECTIVE

The student will recognize terminology used in the metric system.

PROBLEM

Find and circle as many words as you can that have something to do with metric measurements.

NUMBER OF PARTICIPANTS

Any number.

MATERIALS

Teacher-prepared worksheet.

DISCUSSION

How many words can you list that have something to do with metric measurement?

PREREQUISITE SKILLS

Understanding of how to do a word find.

DIRECTIONS

1. How many of the words listed below can you find in Figure 28? The words may be written vertically, horizontally, or diagonally.
2. When you find one of the words in the list, circle it and check it off your list.
3. The first one is done for you.
4. Do any words appear more than once? Which word occurs the most? How many times did it occur?

```
E E R C K K R S L N P O M C Z W L
Q A N T I M N X I V U R O A N K I
U M I L L I M E T E R E E P S O N
I N P R O R N A E B B R A A T S E
V O L U M E B G R A M Q T C S M A
A S E S E E R L K I L O L I T E R
L O N W T X T B R C E N W T S T M
E C G Y E X A R E A B X I Y U V I
N U T Z R I E L I Q U I D N M K L
T B H H E I G H T C C S T R I I L
E E D F G G F H I I S J H R L L I
D E N M P D M E T E R Y D E L O G
I S Q U A R E D R O P E S S I G R
S M I L E S A P A G O O D T M R A
T D E A L F S O T R Y T K Y E A M
A C E L S I U S B H E I E P T M O
N H A N G R A M O S R E C N E D E
C E N T I M E T E R M A R G R N O
E E H E C T O M E T E R W I D T H
```

FIGURE 28.

TEACHING NOTES

Make sure the student understands the rules.

EXTENSIONS

Have students make up their own word finds.

equivalent	volume	length
kilometer	weight	area
meter	kilogram	measure
height	depth	width

Celsius	ten	cube
millimeter	milliliter	liter
gram	capacity	mass
distance	centimeter	squared
milligram	linear	metric system

do you know your metric self?

TEACHING OBJECTIVES

The students will be able to find their measurements using metric measures.

PROBLEM

Determine your metric measurements and put your answers on your "butcher paper" silhouette.

NUMBER OF PARTICIPANTS

Any number.

MATERIALS

Butcher paper, crayons, masking tape, metric tape measure, and a large display area.

DISCUSSION

How well do you know your metric measures?

PREREQUISITE SKILLS

Ability to use several metric measures.

DIRECTIONS

1. Have the students pair off and trace the silhouette of their partner on the butcher paper. This silhouette will become the record book upon which each student records his or her metric measurements.
2. Have the students record on their silhouette the measurements of their:

a. height
b. waist
c. head
d. neck
e. wrist
f. ankle
g. index finger
h. hand span length
i. mass
j. area (decimeter squares)

TEACHING NOTES

These silhouettes make a very nice display for a hallway. Good for an open-house display.

EXTENSIONS

1. Students could determine the distance covered if all silhouettes were laid end to end.
2. Find average mass of all students.
3. Have student cover their silhouettes with decimeter squares and find the approximate area in square meters.

can you fill in the blank?

TEACHING OBJECTIVE

The students will be able to change from one metric unit to another.

PROBLEM

Find the equivalent metric measure for each problem.

NUMBER OF PARTICIPANTS

Any number.

MATERIALS

Prepared worksheet.

DISCUSSION

What relationships exist between units in the customary system? In the metric system?

PREREQUISITES

Knowledge of basic metric conversion factors.

DIRECTIONS

Fill in the missing blanks.

1. 120 mm = _____ cm
2. 280 cm = _____ m
3. 2 km = _____ m
4. 3 l = _____ ml
5. 2.4 l = _____ ml
6. 5 kg = _____ g

7. 5000 mg = _____ kg
8. 6 g = _____ mg
9. 500 ml = _____ l
10. 6.82 t = _____ kg
11. 500 g = _____ kg
12. 150 m = _____ km
13. 0.5 kg = _____ g
14. 0.5 m = _____ mm
15. 0.5 m = _____ cm

TEACHING NOTES

Facility in moving comfortably from one metric unit to another illustrates to the student one of the advantages of the metric system. Make up extra drill problems for students who need the practice.

EXTENSION

Measure objects in your classroom and express them in several equivalent metric units.

do you have it?

TEACHING OBJECTIVE

The student will be able to name the metric units for mass, length, volume, area, and capacity.

PROBLEM

Win the game by completing the most metric books. To do this, you must know your metric units.

NUMBER OF PARTICIPANTS

Two to five.

MATERIALS

Prepared deck of cards.

DISCUSSION

Do you know all of the metric terms?

PREREQUISITE SKILLS

None.

DIRECTIONS

1. The game of "Call It" is played with 20 cards—five books of four cards each. The object is to obtain a complete metric book by calling for cards from other players.
2. Shuffle deck and pass four cards to each player.
3. Place remainder of the deck in the center of the table.
4. Player on dealer's left begins by calling for a specific card from any other

player. If the called upon player has the card, he or she surrenders it to the caller, who continues calling until he or she fails to make a correct call.

5. The caller then draws a card from the top of the deck and play passes to the next player at the left.

6. When a player gets a complete book he or she lays it aside. At the end of the game, the player with the most books wins.

TEACHING NOTES

Using a different colored marker for each book aids in playing the game. A card would look like this:

```
+----------------+
|                |
|   MASS         |
|                |
|   gram         |
|                |
+----------------+
```

The *mass* book would contain milligram, gram, kilogram, and metric ton cards. The *volume* book would contain cubic millimeter, cubic centimeter, cubic meter, and cubic kilometer. The *area* book would contain square millimeter, square centimeter, square meter, and hectare. The *capacity* book would contain milliliter, centiliter, liter, and kiloliter. The *length* book would contain millimeter, centimeter, meter, and kilometer.

For beginners it could be helpful to have the entire book of units listed as below:

```
+----------------+
|                |
|   MASS         |
|                |
|   GRAM         |
|   milligram    |
|   kilogram     |
|   metric ton   |
|                |
+----------------+
```

what is the correct sign?

TEACHING OBJECTIVE

The student will be able to make conversions within the metric system.

PROBLEM

Make each relationship true by placing the correct symbol in each blank.

NUMBER OF PARTICIPANTS

Any number.

MATERIALS

Teacher-prepared worksheet.

DISCUSSION

As you go from one dimensional measure to two and three dimensional measures, what happens to the decimal point?

PREREQUISITE SKILLS

Understanding of relationships between metric units (subdivision and multiples).

DIRECTIONS

Place the $<$, $>$, or = symbol in the blank space to make the relationship true.

1. 1 kg _____ 100 g
2. 1000 dm^3 _____ 1 m^3
3. 1 cm _____ 1 mm

4. 100 mm _____ 1 cm
5. 1000 cm _____ 1 m
6. 100 g _____ 1 kg
7. 1 liter _____ 1 dm^3 (water)
8. 1 cm^3 _____ 10 ml (water)
9. 10 ml _____ 1 liter
10. 10 kg _____ 1 metric ton
11. 1 m _____ 1000 mm
12. 1 000 000 cm^3 _____ m^3
13. 1000 mm^3 _____ 1 cm^3
14. 1000 m^2 _____ 1 hectare
15. 10 000 m^2 _____ 1 km^2

TEACHING NOTES

Students may write in the words "more than," "equals," and "less than," if they have problems with the relationship symbols.

EXTENSION

Have students make up problems and exchange papers with their classmates.

what would it be like living metrically?

TEACHING OBJECTIVES

The students will be able to find a number of everyday events that can be measured using metric measures.

PROBLEM

Spend one day "metrically," use only metric measurements on this day.

NUMBER OF PARTICIPANTS

Any number.

MATERIALS

None.

DISCUSSION

What activities in our daily lives are measured?
Are our measures accurate?
Do we use any measurement sayings in our daily lives?

PREREQUISITE SKILLS

Good knowledge of measurement.

DIRECTIONS

Set aside a "Metric Day" where you will record in metric units everything that you come in contact with that will be changed because America is going metric. Don't forget the little things you use every day.

TEACHING NOTES

This may take practice to get yourself and students aware of the many areas affected by measurement.

EXTENSIONS

Make a bulletin board showing the things that will change as the United States changes to the metric system.

where is the decimal? what is the word?

TEACHING OBJECTIVE

The student will be able to determine the correct metric word and the correct placement of the decimal so that sentences make sense.

PROBLEM

Make each sentence metrically correct by putting in the proper word or placing the decimal point in the correct place.

NUMBER OF PARTICIPANTS

Any number.

MATERIALS

Prepared worksheet.

DISCUSSION

Why are decimal points important?

PREREQUISITE SKILLS

Knowledge of metric measuring units.
Understanding of correct placement of decimal point.

DIRECTIONS

Below are a number of statements in which either the decimal point or the proper measurement word is missing. Fill in the proper word and/or put in a decimal point so that these statements are metrically correct.

1. For the holidays a 450-kg turkey was purchased.
2. If you are healthy, your temperature would be 370°
   _____.
3. They needed 270 _____ of material to make the dress.
4. The posted speed limit was 55 _____ per hour within the
   city limits.
5. The water started to boil when the temperature went to
   1 000 000°C.
6. The milkman delivered 2000 liters of milk to our house
   daily.
7. The student weighed 498 kg.
8. The average mass of the five women was 450 kg.
9. We baked the cookies for 15 minutes at 1700° _____.
10. His average speed was a little over 800 km/hr.
11. The room measured 36 500 m by 50 400 m.
12. Her car took 760 liters of gas.
13. The bottle held 4000 ml of pop.
14. The tallest student in the class was 173 m tall.

TEACHING NOTES

Have "approximate" conversion charts available for those students who
need this resource.

EXTENSION

Additional exercises could be written by students following the same
format to test their knowledge of our metric world.

how much does it cost?

TEACHING OBJECTIVE

The student will be able to relate metric measures to prices.

PROBLEM

Be the first one in your class to find the correct prices using the information given.

NUMBER OF PARTICIPANTS

Any number.

MATERIALS

Prepared worksheet.

DISCUSSION

As you buy things by unit prices, how will metric measurements affect the prices?

PREREQUISITE SKILLS

Ability to multiply fractions, decimals, and whole numbers.

DIRECTIONS

In the following tables find the price on the items from the given information.

TEACHING NOTES

This will probably require additional help for most students.

Given	Amount Bought	Total Price
Hamburger	0.5 kg	_____
sale:	1.5 kg	_____
$2.50/kg	2000 g	_____
	1.2 kg	_____

Given	Amount Bought	Total Price
Fabric sale:	6.25 m	_____
wool	3.75 m	_____
$3.00/m	1500 cm	_____
	7.5 m	_____

Given	Amount Bought	Total Price
Tile squares:	3 m X 3 m	_____
11¢ per 30 cm	1 m²	_____
by 30 cm tile	30 m²	_____
	3 m X 6 m	_____

Given	Amount Bought	Total Price
Gasoline:	10.2 l	_____
15¢ per liter	23 l	_____
	5.7 l	_____
	40.4 l	_____

Given	Amount Bought	Total Price
Storage space	2 m X 6 m X 4 m	_____
for rent:	3 m on a side	_____
20¢ per cubic	1.5 m on a side	_____
meter per month	0.5 m X 0.25 m X 5 m	_____

EXTENSIONS

Have students make up their own product advertisements.

can you find the pattern?

TEACHING OBJECTIVE

The student will be able to make conversions within the metric system and to find patterns and complete them.

PROBLEM

Determine the pattern and then complete the charts.

NUMBER OF PARTICIPANTS

Any number.

MATERIALS

Prepared worksheet.

DISCUSSION

What operations could you use to help determine a pattern?

PREREQUISITE SKILLS

Ability to add, subtract, multiply, and divide.

DIRECTIONS

Find the pattern and fill in the blank spaces:

30 cm	1.2 m
500 ml	2 l
250 g	1 kg
5 mm	_____ cm
1.5 cm	_____ cm
50°C	_____ °C

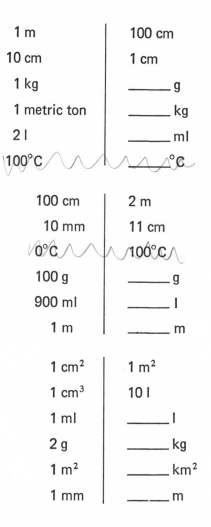

1 m	100 cm
10 cm	1 cm
1 kg	_____ g
1 metric ton	_____ kg
2 l	_____ ml
100°C	_____°C

100 cm	2 m
10 mm	11 cm
0°C	100°C
100 g	_____ g
900 ml	_____ l
1 m	_____ m

1 cm²	1 m²
1 cm³	10 l
1 ml	_____ l
2 g	_____ kg
1 m²	_____ km²
1 mm	_____ m

TEACHING NOTES

Have students convert to a common unit to find the pattern.

EXTENSIONS

Have students make up similar exercises.

what will the weather be?[1]

TEACHING OBJECTIVE

The students will be able to use metric measures to record the weather.

PROBLEM

Report the weather using only metric measurements.

NUMBER OF PARTICIPANTS

Any number.

MATERIALS

Weather report from newspaper, radio, or television.

DISCUSSION

What information given in a weather report is a measurement?

PREREQUISITE

Ability to use conversion formulas.

DIRECTIONS

1. Bring in from the newspaper the section on weather reporting.
2. Determine the weather in degrees Celsius.
3. Determine wind velocity using metric measures.
4. Determine rainfall using metric measures.
5. Write a metric weather forecast.

---

[1] This activity requires conversions.

TEACHING NOTES

This is one activity that conversion between customary and metric will have to be made. This activity should be omitted if you are opposed to any general conversion.

EXTENSION

Make a metric weather bulletin board for your school and change it daily.

7

Ideas

Included in this chapter are various change of pace activities.

Five-minute fillers are included for all major areas of metric measurement. These fillers are available for those short periods of class time that are long enough to be productive yet too short to complete an entire lesson. The entire class can work on these activities or they can be assigned to the student to complete on his or her own free time. The five-minute fillers are short, interesting, and require no preparation on the teacher's part.

The class activities are not divided into topics, but cover all areas of metric measurement. These activities are intended to involve an entire class or group. Some of these can be accomplished at a single sitting, while others will require a longer period of time. Generally they include games, activities, and contests.

Some sample bulletin boards are also included. These are ideas which, hopefully, will help the teacher to teach, as well as generate enthusiasm, interest, and excitement over the metric system.

FIVE-MINUTE FILLERS

linear

1. How many words can you make from the letters in LINEAR MEA-SURE? *Note:* Each letter may be used only as many times as it appears. Example: Lineal cannot be used as there is only one "l" in "linear measure."
2. How many items can you list that are approximately 1 m long?
3. How many items can you list that are approximately 10 cm long?
4. How many items can you list that are less than 1 m long?
5. Say and Draw
   a. One child is sent out of the room.
   b. While the child is out of the room the teacher (using a metric ruler) draws a figure on the board. (as in Figure 29.)
   c. The class sketches the drawing (putting down the length in metrics of each line segment drawn).
   d. The figure is then covered up.
   e. The student returns to class and asks the assistance of his or her classmates in making the figure.
   f. The student would ask one of his or her classmates how long to draw the first line segment. He or she then would ask another student where and how long to draw the second line. The student proceeds until the drawing is complete.
   g. Now compare the two drawings.

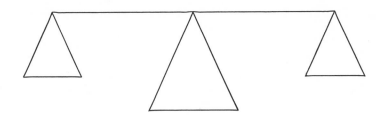

FIGURE 29.

6. Measure the diameter of a half-dollar, a quarter, a dime, a nickel, and a penny.

7. What's My Name?
   a. On 5 cm X 10 cm cards have students write kilometer, decimeter, meter, centimeter, millimeter. One word to a card.
   b. The teacher would hold up a symbol of one of the metric linear terms and each student would show the correct word.

8. Pick a cartoon character and measure metrically the length of arm, fingers, feet, body, etc.

9. Write a limerick, ad, poem, riddle, story, etc. about linear measures.

area and perimeter

1. Find area (perimeter) of your reading book, teacher's desk, your desk, table, etc.

2. Figure out the approximate area (perimeter) of a cartoon character, picture, leaf, etc.

3. Find the total area of all the notebook paper you use each day. What would be the total area of notebook paper the entire class uses each day?

4. What would be the total area of notebook paper the class uses for one assignment?

5. Find the area (perimeter) of various size television screens.

6. How many items can you list that you would measure in square meters?

7. How many items can you list that you would measure in square centimeters?

8. How many items can you list that you would measure in square millimeters?

9. How many items can you list that you would measure in square kilometers?

10. Write a limerick, ad, poem, riddle, story, etc. about area (perimeter).

volume and capacity

1. How many words can you make from the letters in LIQUID CAPACITY? *Note:* Each letter may be used only as many times as it appears.

2. How many items can you list that will be sold by the liter?

3. From a magazine cut out pictures of items that will be sold by the liter and make a collage.

4. Suppose the classroom was full of Kool Aid. How much Kool Aid would there be?

5. Figure out the volume of water you drink on one day.

6. Write a limerick, ad, poem, riddle, story, etc. about the liter.

7. Make up cards labeled as follows:
   1000 ml   100 ml   10 ml   1000 cm^3
   one liter   1 m^3 10 cm^3 100 cm
   60 ml   100 cubic centimeters
   Now have students pick up cards which are names for 1 liter.

8. Find the capacity of the classroom aquarium.

9. Find the volume of a dictionary, an encyclopedia, a set of encyclopedias.

mass

1. How many items can you list that are now measured in grams?

2. How many items can you list that have a mass less than 1 g?

3. Make a collage of pictures of items that will have a mass more than 1 g but less than 10 g. Pictures can be obtained from magazines and newspapers.

4. Describe the difference between weight and mass.

5. Make a collage of items that will have a mass of less than 1 g. Pictures may be obtained from magazines and newspapers.

6. Write a limerick, ad, poem, riddle, story, etc. about mass.

7. Find the masses of various animals.

8. Place five objects in front of the room. Have a student decide which is the heaviest and lightest. Have the student order them from lightest to heaviest.

9. Show students objects with the correct mass labeled on the object. Remove label with correct mass and have students come up and place correct label with correct object.

temperature

1. How many words can you make from the letters in CELSIUS THERMOMETER. *Note:* Each letter may be used only as often as it is found in the words.

2. How many items can you list that will be affected by changing to degrees Celsius?

3. Write a limerick, ad, poem, riddle, story, etc. about degrees Celsius.

4. Pick a short range of Celsius temperatures and do a collage illustrating temperatures within the chosen range. Pictures can be obtained from magazines and newspapers.

5. Keep track of the temperature in degrees Celsius for a two-week period. Find the average Celsius temperature for this period.

6. Fill a glass with drinking water, place glass outdoors, and record temperature every hour. Fill another glass with drinking water, place glass in classroom, and record temperature every hour. Compare the results.

CLASS ACTIVITIES

1. Design a metric T-shirt.
2. Design a metric poster.
3. Design a metric bike bumper sticker.
4. Write a limerick, ad, poem, riddle, story, saying, etc. dealing with metrics.
5. Rewrite some popular sayings, changing to metrics wherever necessary.
   a. An ounce of prevention is worth a pound of cure.
   b. I wouldn't touch it with a ten-foot pole.
   c. There was a crooked man who walked a crooked mile.
   Have students find other sayings that would be changed.
6. Make a graph comparing heights of students in metric measure.
7. Make a graph comparing the mass of students.
8. Conduct a "metric Olympics."
   a. straw javelin
   b. paper-plate discus
   c. broad jump
   Students could list other categories.
9. Have a metric scavenger hunt. For example, find an object that has a mass of 2 kg, find an object that is 5 cm long, etc.
10. On the playground measure off 1 m (using trundle wheel) and place a stake at each end of the meter. Connect stakes with taut string. Have babysteps races, jumping races, etc.
11. If space permits, have 25-, 50-, and 100-m races.

12. Have a contest to estimate length.
    a. Divide class into equal-ability groups.
    b. Give each group a list of measurable objects that are in the room, e.g., length of chalk board, length of eraser, etc.
    c. Each group estimates the length of each object on their list.
    d. Groups exchange papers and students actually measure items on list and put down their findings.
    e. Return papers to their owners.
    f. Group that was closest to actual measurement for the first object gets three points for their team.
    g. Proceed for each object as in step f.
    h. Winning team is the team with the most points.
13. In five minutes list as many metric terms, symbols, etc. as you can that have to do with the liter, gram, meter, or degrees Celsius.
14. Find as many items as you can in the newspaper that will be affected by changing to the metric system.
15. Find a food ad and change the ad to metrics.
16. Visit a grocery store and see how many different metric units are used on labeling.
17. Collect car brochures and see how many different metric units are used in car specifications.
18. Invite in metric-resource people from city or county government and have them explain the effects of the metric system on the community.
19. Set up a metric football field or basketball court and change the rules of the game using correct metric measurements.
20. Make a "weather booklet" using only metric terminology.
21. Change a recipe to metrics and try it out.
22. Make a metric mobile.
23. Make puzzles (Figure 30) using metric symbols, terms and equivalents. Each puzzle piece should fit together uniquely.
24. Do science experiments and charts using metric measures.
25. Chart the growth of a seed metrically.
26. Play "20 Questions" to identify objects in the room. Each question must have at least one metric term in it.
27. Bring in items that have dual measurements listed (customary and metric).
28. Write metric words on 5 cm × 10 cm cards. One word to a card. Have

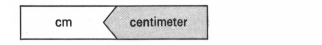

FIGURE 30.

      students match cards to various categories. For example, all cards that have to do with measuring length in one group, etc.

29. Write on the blackboard metric terminology and symbols. Have students match the correct word with the correct symbol.

30. Design or lay out a metric city.

BULLETIN BOARDS

Monkeying Around

How long are the tails?

A–F G–L M–R S–Z

Library-card pockets

To the teacher: Students will measure tails of monkeys, put their answers and their names on a 3 × 5 card and place the card in the pocket according to their last name.

Who-o-o-o-o

Can make these metric unit conversions

?

1 cm = _____ mm

1 ml = _____ l

1 kg = _____ g

1 m = _____ cm

1 mm = _____ cm

1 km = _____ m

1 t = _____ kg

Let's Talk Metrically

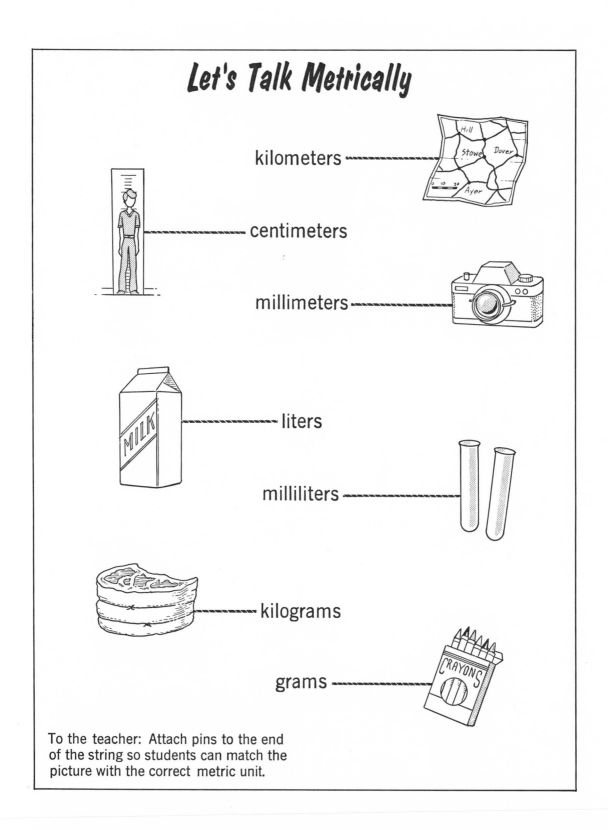

kilometers

centimeters

millimeters

liters

milliliters

kilograms

grams

To the teacher: Attach pins to the end
of the string so students can match the
picture with the correct metric unit.

THE CELSIUS WAY

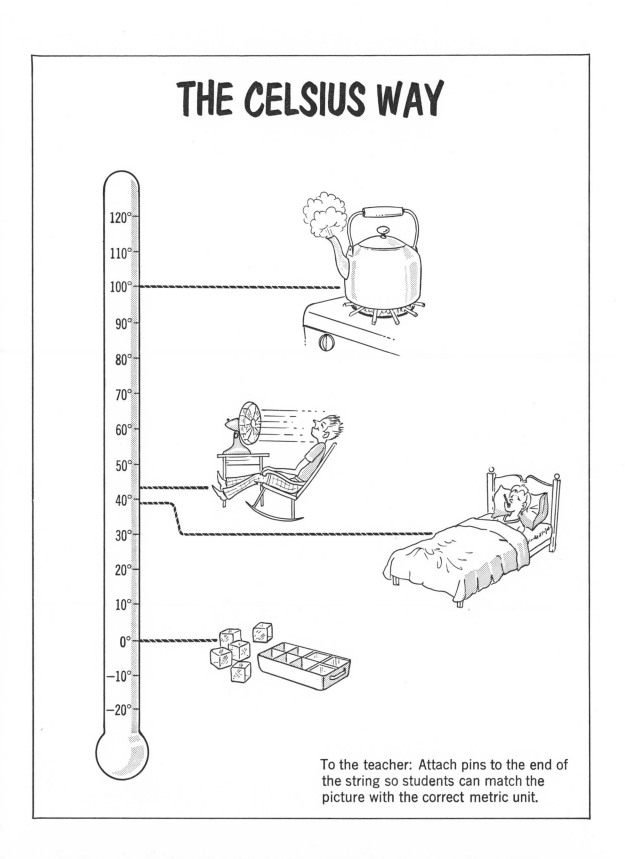

To the teacher: Attach pins to the end of the string so students can match the picture with the correct metric unit.

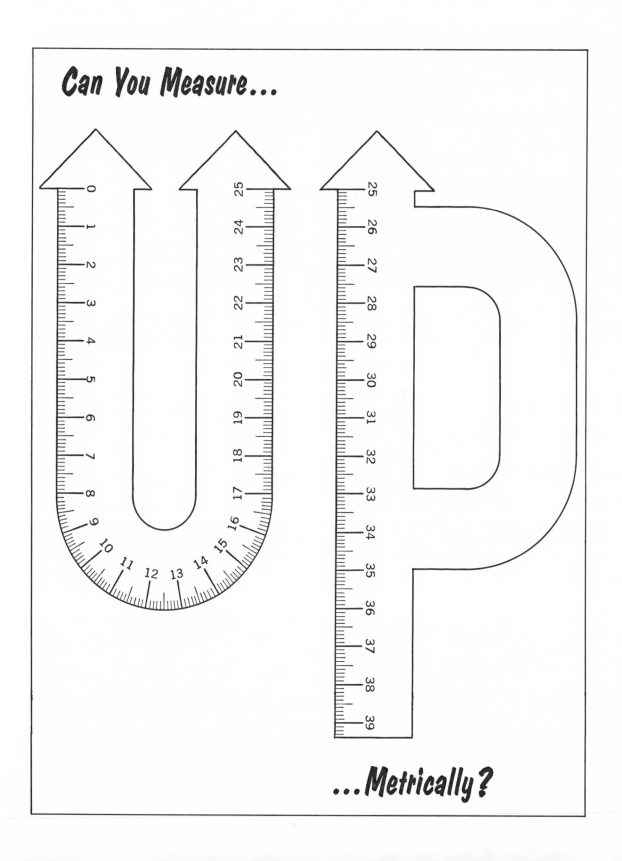

How Many Will It Take?

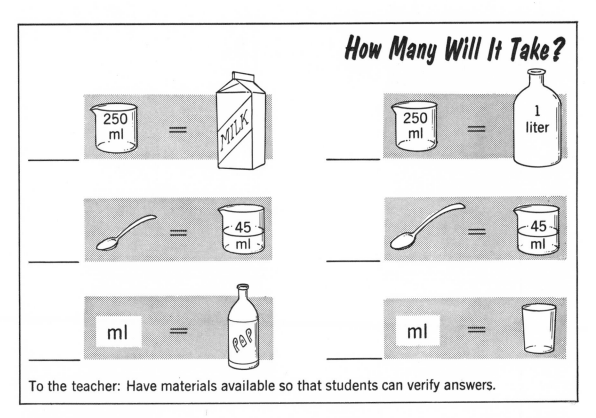

_____ 250 ml = MILK

_____ 250 ml = 1 liter

_____ spoon = 45 ml

_____ spoon = 45 ml

_____ ml = POP

_____ ml = glass

To the teacher: Have materials available so that students can verify answers.

Grampa Says...

1 liter = _____ milliliters

1 kilogram = _____ grams

1 meter = _____ centimeters

1 meter = _____ millimeters

To the teacher: Strips can be changed to teach different equalities

Balance Me

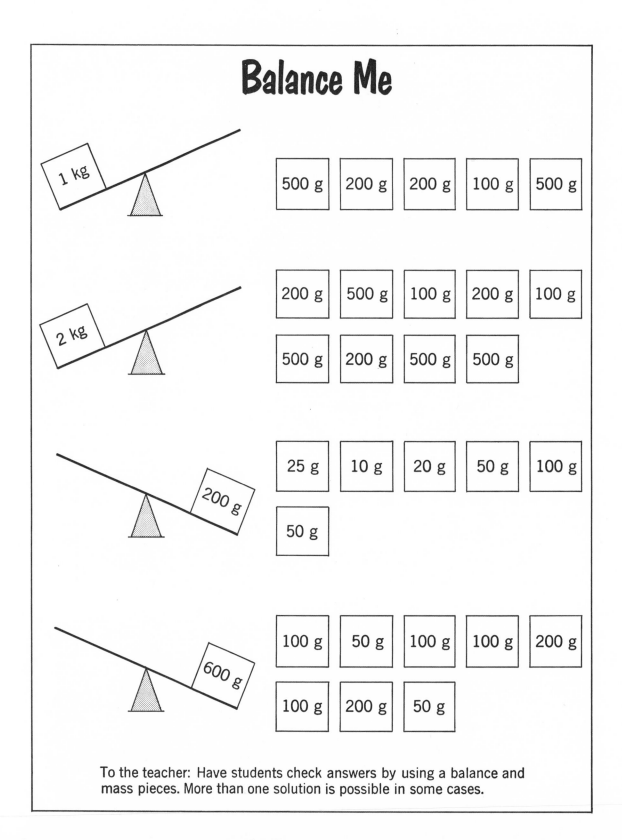

To the teacher: Have students check answers by using a balance and mass pieces. More than one solution is possible in some cases.

A Metric Zoo

How long is
the trunk?

How long is
the neck?

How long is
the tail?

How long is
the ear?

To the teacher: Students will measure trunk, neck, tail and ear of animals, place their answers with their names on a 3 × 5 card and place in correct library pocket.

What Should I Wear?

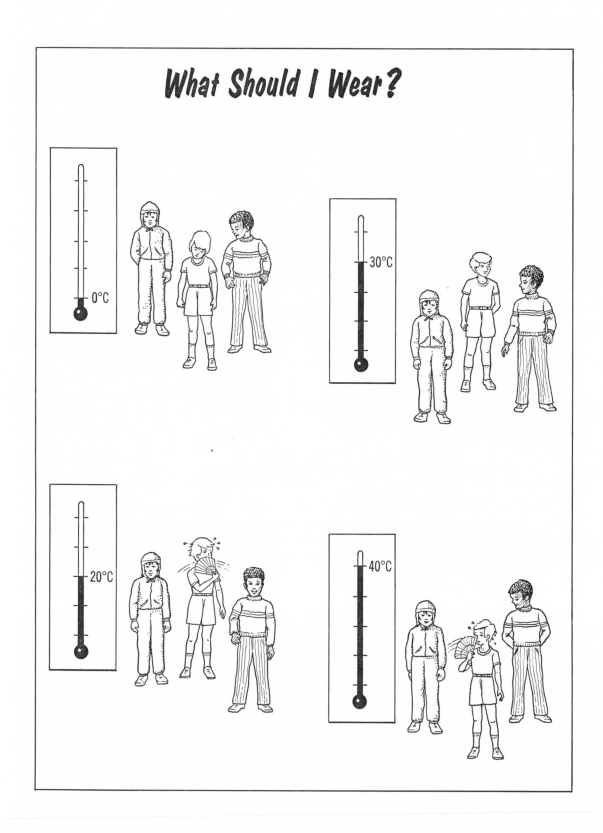

The Metric System in the United States

"Measures and weights," observed John Quincy Adams, "are the instruments used by man for the comparison of quantities and proportions of things."

The ancient units of measurement were often descriptive of that which served as the standard and also exhibited a naturally occurring ratio between the multiples and subdivisions. Ethnic conditions, the whims and caprices of rulers, imposition and fraud, conquest, and methods and habits of thought and life: all have had their effect on the development and use of measuring systems. However, by the middle of the nineteenth century two systems of weights and measures had attained predominance throughout the world—the English/American customary system and the so-called French metric system.

The contrast in the ways in which the British and French went about solving the confusion in weights and measures is noteworthy. France simply discarded its old system and substituted a new one in its place. England, on the other hand, gradually improved its system of weights and measures through the enforcement of stricter laws. Only in recent years has England felt enough pressure to "go metric" along with most of the rest of the world. The French system grew because of its appealing simplicity and logic. The English system was the basis for the vast majority of commercial dealings and, even more important, was the one used to construct British and American machinery, which was much in demand.

Until the middle of the twentieth century, then, most of the English-speaking countries of the world retained the customary system of weights and measures.

The metric system was created whole and put into use under unusual conditions and to serve very specific purposes. It was based on what were, at the time of its creation, the most advanced scientific principles known. Although the intellectual foundation for the metric system had been laid by the rebirth of scientific interest in France between the sixteenth and eighteenth centuries, the French Revolution was the catalyst which propelled it into a practical reality.

In searching for a "founding father" of the metric system, historians have settled on Gabriel Mouton (1618–1694), the vicar of St. Paul's Church in Lyons, who proposed a comprehensive decimal system of weights and measures in the year 1670.

In 1790, the French Academy appointed several committees to investigate weights and measures reform. One of these committees reported on October 27, 1790, urging the adoption of a decimal basis for the new system. After consideration of several alternative possibilities, the committee recommended the adoption of a unit equal to one ten-millionth of the length of a quadrant of the earth's meridian (i.e., one ten-millionth of an arc representing the distance between the equator and the north pole). The unit of mass was to be derived by cubing some part of this length unit and filling it with water. The same technique would also provide the capacity measure. In this way, the standards of length, mass, and capacity were all to be derived from a single measurement, infinitely reproducible because of natural origins, precisely interrelated, and decimally based for convenience.

It cannot be claimed that the new system met with instantaneous approval. Even in France the transition was not effected quickly or smoothly, for although the system was made mandatory throughout France in 1795, its use was not enforced. The acceptance of the metric system suffered another blow in 1812, when Napoleon Bonaparte issued a decree allowing the old units to return to use. For a time, Napoleon's act was popular, probably because the metric system had not been in use long enough to fully supplant the older and more familiar system. Then, in July of 1837, the following act was passed: "After January 1, 1840, all weights and measures other than the weights and measures established by the laws of 1795 and 1800, constituting the decimal metric system, shall be forbidden."

Following this action, the metric system began to experience a gradual but steady growth which saw it taken up by one country after another. In fact, the growth in the use of the metric system after 1850 is little less than phenomenal. By 1880 seventeen nations—including most of South America and the major European nations of Germany, Austria-Hungary, Italy, and Norway—had officially accepted the metric system at least for government purposes, and eighteen more nations were added to this list by 1900.

Among the leading industrial nations of the world at the dawn of the twentieth century, only Britain and the United States had not accepted the metric system.

From its very beginnings as an independent nation, the United States has been concerned with its system of weights and measures and has often considered the notion of changing it. The first attempt came at about the same time that a complete decimal system of coinage was approved by Congress on August 8, 1786. Thomas Jefferson proposed both the decimal system of coinage and a decimal system of weights and measures. He considered the decimal system of weights and measures to be a logical extension of the decimal coinage concept which was adopted by Congress. Jefferson never became an advocate of the metric system because the basic measurement could not be reproduced in any country except France, so that other nations would either have to trust the French results or take the trouble to send people to France to verify it for themselves. He wanted the length standard to be based on a one-second arc of a pendulum at a latitude of 45 degrees. His recommendation was not passed by Congress.

John Quincy Adams submitted a comprehensive report in 1821. This report explained the concepts and practices of weights and measures. It also listed the advantages and disadvantages for the United States of both the English and metric systems. Adams did not believe that the metric system had yet attained sufficient maturity to recommend its adoption by the United States in 1821.

Prior to the American Civil War, not one piece of legislation calling for acceptance of the metric system had even been introduced in Congress. Then, in 1866, an act was passed without resistance or fanfare, making it legal to use the system for the transaction of any and all business in this country.

Events which led to the Act of 1866 included the adoption of the metric system by several European countries and its facility for scientific work. By 1850 an acute need was felt for a universal language of science. These events led Senator Charles Sumner of Massachusetts to comment.

"By these enactments the metric system will be presented to the American people, and will become an approved instrument of commerce. It will not be forced into use, but will be left for the present to its own intrinsic merits. Meanwhile it must be taught in schools. Our arithmetics must explain it. They who have already passed a certain period of life may not adopt it; but the rising generation will embrace it and ever afterwards number it among the choicest possessions of an advanced civilization."

Because the Act of 1866 had not made the metric system mandatory for the people of the United States, a great deal of missionary work remained to be done if the objectives set by the Committee of Coinage, Weights, and Measures were to be accomplished. Before long, a full-fledged promotional campaign in favor of metric usage had been established. This in turn generated the first American countermovement in opposition to its adoption.

The leader of the countermovement to metric adoption was the International Institute. The apex of the Institute's racially prejudiced, antimetric stance was reached in 1883 with the composition and publication of its theme song, entitled "A Pint's a Pound the World Around." Two of its more colorful stanzas and the chorus went as follows:

They bid us change the ancient "names,"
  The "seasons" and the "times;"
And for our measures go abroad
  To strange and distant climes.
But we'll abide by things long clear
  And cling to things of yore,
For the Anglo-Saxon race shall rule
  The earth from shore to shore.

Then down with every "metric" scheme

Taught by the foreign school,
  We'll worship still our Father's God!
    And keep our Father's "rule"!
A perfect inch, a perfect pint,
  The Anglo's honest pound.
Shall hold their place upon the earth,
  Till Time's last trump shall sound!

CHORUS:

Then swell the chorus heartily,
  Let every Saxon sing:
"A pint's a pound the world around,"
  Till all the earth shall ring,
"A pint's a pound the world around,"
  For rich and poor the same;
Just measure and a perfect weight
  Called by their ancient name!

Building upon a mystical explanation of the metrological revelations of ancient Egypt's "Great Pyramid," the International Institute successfully fought the metric system on the grounds that it was neither God-given nor Anglo-Saxon, two unpardonable attributes.

Only a few years elapsed between the waning of the intense interest in the question of metric adoption in the 1880s and the time when the issue was revived. After an announcement by the Treasury Department in 1893 that the nation's "fundamental standards" would thereafter be metric, and following the inclusion of metric system provisions in weights and measures laws passed in 1893 and 1894, efforts to effect greater use of the system through congressional action were stepped up. In 1896 the House passed a bill that would have achieved the long sought goal, but then voted to reconsider its action and finally sent the measure back to the Committee on Coinage, Weights, and Measures. This was as close as the metric system was to come to achieving legislative endorsement for many years in this country.

Between 1888 and 1902 the basic assumption that adoption of the metric system was inevitable and advantageous had gone largely unchallenged. It was true that Congress had never been able to legislate an exact date for the changeover to occur, but most participants in the debate had agreed that, sooner or later, it would have to happen. Beginning in 1902, however, a chorus of dissent arose that was to grow ever louder as time passed.

The next metric campaign was launched in 1916, blossomed after the armistice, reached peaks of furious activity in 1921-1922, and 1925-1926, and burned itself out in the early years of the depression. During these years the metric issue became a full-fledged public controversy. Following the 1926 failure, the great metric crusade slowly began to atrophy. Although occasional spurts of activity continued to occur until 1933, none of these represented much more than a half-hearted attempt to revive an issue which nearly everyone admitted was dead. Nearly a quarter of a century elapsed before anything even approaching widespread interest in the metric system developed again. A depressed American economy and an isolationist political mood were primarily responsible for this lack of interest in going metric.

In late 1958 the British Association for the Advancement of Science launched another investigation into the metric system to attempt to find out what it would cost Great Britain to change and what the long-term benefits of metric adoption would be. Since that time Great Britain has made the change. Early in 1959 the American Association for the Advancement of Science followed suit by establishing a committee to consider the problems involved in a change after approving in principle the general adoption of the metric system.

In the late sixties and early seventies, bills to force the United States to change to the metric system have been introduced to the Congress, but passage was continually blocked.

metric historical dates

1670— Gabriel Mouton, Vicar of St. Paul's Church in Lyons, France proposed a comprehensive decimal system of weights and measures that was based on a unit from the physical universe instead of the human body.

1786— A complete decimal system of coinage was adopted for the United States.

1790— Thomas Jefferson submitted a report on weights and measures to Congress. A basic standard, derived from the motion of the earth on its axis was proposed to establish a decimal system of weights and measures.

A French decree led to the development of the metric system.

1795—	A French decree was issued officially adopting the metric system. Copies of the provisional standards were sent to several countries, including the United States.
1799—	The first federal weights and measures law was enacted.
1812—	By decree, Napoleon Bonaparte temporarily suspended the compulsory provisions of the 1795 metric system law.
1821—	Secretary of State John Quincy Adams submitted an exhaustive report on the subject of weights and measures to Congress in response to a resolution passed by the Senate in 1817. Adams recommended retention of the English customary system by the United States, but he proposed a program for achieving greater uniformity among the states.
1832—	By administrative action, the Secretary of the Treasury declared the yard, the avoirdupois pound, and the Winchester bushel to be the official system of weights and measures for the United States.
1837—	The metric system was made compulsory again in France.
1840—	Greece, the Netherlands, and Italy went metric.
1849—	Spain went metric.
1866—	Use of the metric system was made legal in the United States by an Act of Congress.
1868—	Portugal went metric.
1873—	The American Metrological Society was organized in New York for the purpose of improving existing systems of weights, measures, and moneys.
1875—	The Treaty of the Meter was signed in Paris by seventeen nations, including the United States.
1880—	Most of Europe and South America went metric.
1893—	The Superintendent of Weights and Measures issued a bulletin announcing that the U.S. prototype meter and kilogram would henceforth be considered the nation's "fundamental standards of length and mass."
1894—	A law defining and establishing units for electrical measurement was passed by Congress. These units were based on the metric system.
1895—	A resolution establishing a commission to study and report on the feasibility of metric adoption was passed by the House of Representatives. By mistake, the resolution was recorded as requiring the concurrence of the Senate in order to be put into effect. Consequently, the commission was never formally organized.

1897— Legislation was enacted by Great Britain permitting full use of the metric system.

1907— Following a refusal by the Committee on Coinage, Weights, and Measures to report favorably on a metric bill, intense promotional efforts died down until the advent of World War I.

1918— General Order No. 1 was issued by the War Department provided for the usage of the metric system for wartime activities.

1937— A bill to fix the standards according to the metric system was considered and recommended by the Committee on Coinage, Weights, and Measures. But the bill was never enacted.

1948— The "Twentieth Yearbook of the National Council of Teachers of Mathematics" was devoted solely to a discussion of the need for and advantages of using the metric system, particularly for educational purposes.

1957— The U.S. Army issued a regulation establishing metric linear units as the basis for weapons and related equipment.

1960— At the 11th General Conference on Weights and Measures a new international standard of length, based on the wavelength of the element krypton was adopted in place of the original "meter bar." At the same conference, the modernized metric system was officially renamed the Systéme International d' Unités—The International System of Units.

1965— Britain started its 10-year metrication program.

1968— A study providing for a three-year program to determine the impact of increasing use of the metric system in the United States was passed by Congress and signed into law by President Lyndon B. Johnson.

1970— Canada issued a statement which said that metrication is a definite objective of Canadian policy.

1972— The U.S. Senate unanimously passed the "Metric Conversion Act of 1972."

1974— The U.S. House of Representatives in a 240 to 153 vote defeated a motion to suspend the rules to consider metric conversion legislation (H.R. 11035) without any amendments being attached.

reference

U.S. Metric Study Interim Report: A History of the Metric Controversy in the United States. Washington: Government Printing Office, 1971.

9

Description of Derived Units

Measurement is an important bridge between science and mathematics. Science poses measurement problems and mathematics provides the tools used to solve them. Metric measurement, which has long been the tool of science, is now becoming the measurement system of mathematics. The following science measurements illustrate the application of the metric system to the world of science. To express measurements in everyday life as well as in the sciences, a combination of measures is used.

speed

When you are traveling in a car or plane, or even walking, you measure the speed with which you are moving in terms of length and time. Speed is the distance an object has moved divided by the time it took the object to move that distance. A baseball travels about 18 m from the pitcher's hand to the catcher's mitt. If it were determined that the baseball traveled the 18 m in 0.5 seconds, we could calculate that the speed of the ball was 36 m/s. If you traveled 85 km in one hour in your car, your speed would be 85 km/h.

velocity

Velocity and speed are often used synonymously; however, there is an important difference between speed and velocity. Speed indicates the rate of motion in any direction. But velocity indicates the rate of motion in one fixed direction. The units used to express velocity are the same ones used to express measurements of speed.

acceleration

Acceleration is the change of velocity of an object divided by the time during which the change takes place. If a car started from rest and in 10 seconds it had reached a velocity of 50 m/s, its acceleration would be: acceleration = change of velocity/time interval of $a = (50$ m/s $- 0.0$ m/s$)/ 10$ s. We would say the acceleration is "five meters per second per second." Acceleration units are velocity units divided by time units, e.g., (km/s) /s. Often an acceleration of 5 m/s /s is written 5 m/s^2.

force

A force is a push or pull exerted on a body. Often force is defined as any quantity that is capable of producing motion. To produce motion there must be an unbalanced force acting on a body. Force is measured in newtons. A newton is measured in kilogram-meters per second squared.

work

In science, work has a precise definition. Work is the product of the force applied to an object and the distance the object moves. In the metric system work is measured in joules. A joule is defined as newton-meters, or kilogram-meter squared per second squared.

Power is the amount of work done per unit of time. The more rapidly work is done, the greater is the power. Average power = work done/time taken to do work. Power is measured in joules per second, or watts. A watt is defined as a joule per second. In the metric system the commonly used term "horsepower" is replaced by kilowatts. A kilowatt is equal to 1000 watts.

The energy of a body is its ability to do work. Since the energy of a body is measured in terms of the work it can do, it has the same units as work.

Pressure is defined as force per unit area. The units used in measuring pressure are newtons per square meter.

Density is the mass of a substance per unit volume. Since volume is usually measured in cubic meters or cubic centimeters, the measurement of density is expressed in kilograms per cubic meter or grams per cubic centimeter. For example, the density of mercury is 13.5 g/cm^3, the density of water is 1.00 g/cm^3, and the density of lead is 11 g/cm^3.

Specific gravity is the ratio of the weight of an object to the weight of an equal volume of water. For example, aluminum has a weight density of 2.7 g/cm^3 and a specific gravity of 2.7 g/cm^3/1.0 g/cm^3 = 2.7. Note that the measure of specific gravity does not have units because it is a ratio of two like quantities.

mole

The mole is a unit used in chemistry to measure amounts of chemicals that take part in chemical reactions. One mole of a substance contains 6.02×10^{23} atoms, molecules, ions, or radicals. The weight in grams of one mole of any substance is the same as the substance's formula weight. For example, one mole of oxygen would weigh 32 g, and a mole of water would weigh 18 g.

degrees kelvin

The temperature $-273°C$ is called the absolute zero of temperature. At this temperature molecules of an ideal gas would cease to move, according to the kinetic theory. The temperature scale that uses $-273°C$ as its zero is called the Kelvin scale, named for the English physicist Lord Kelvin. This temperature is used when working with the laws involving the reaction of gases to pressure, volume, and temperature changes.

heat

Heat is a form of energy. The unit used in the metric system to measure the quantity of heat is the joule—the unit of work or energy equal to 10^7 ergs. Another common unit for measuring heat is the calorie, which is defined as 4.184 joules.

light

Light is measured in lumens. One lumen is the amount of luminous flux incident upon a spherical surface of area 1 m² at a distance of 1 m from a uniform source of one candela. The lumen is a unit of power rather than a measure of energy. A source having a luminous intensity of 1 candela in all directions radiates a light flux of 4π lumens.

10

Careers and Metric

Changing to the metric system affects almost every occupation in one way or another. Temperatures, distances, and weights all change in units of measure under the metric system. The following is a listing of a few occupations and the change which the metric system brings.

1. Education
   Customary U.S. measurement terms will change.
   Less time spent on teaching fractions in the elementary school and more on decimals and power-of-10 notation.
2. Automobile manufacturing
   Types and measures of fasteners will change.
   Engine sizes and measurements.
   Speedometer to kilometers per hour.
   Odometer to kilometers.

3. Engineers

Many already use metric measure but common materials will be measured in metric units.

4. Physicians

Already use metric.

5. Pharmacists

Already use metric.

6. Foresters

Land area will be measured in square kilometers or hectares.

7. Real estate sales

Land area will be measured in hectares, square kilometers, or square meters.

8. Plumbers

Different pipe and thread sizes.

9. Optometrists

Already use metric.

10. Geographers

Measure land area in square kilometers.

11. Journalism

Common U.S. customary measurement terms used in writing will be changed to metric.

Metric paper sizes.

12. Fire fighters

Miles replaced with kilometers.

Liquids measured in liters and kiloliters.

13. Surveyors

Change to metric distance measures.

Instruments will be calibrated in metric measures.

14. Cooks and chefs

Quantities measured in grams, kilograms, liters, and milliliters.

Recipes changed to metric units.

Oven temperatures to use degrees Celsius.

Utensil sizes will change.

15. Carpenters

Linear measure units will change.

Computations will be easier.

16. Gasoline service attendents

Products labeled in metric units.

Mileage computed as km/l or kilometers per liter.

17. Electricians

Already use some metric units.

Measurements in metric units.

18. Farmers

Land measure in hectares.

Crop yield quantities in metric units.
19. Mining
Land measure.
Quantities in metric units, e.g., metric tons.
20. Manufacturing
Metric tools.
Metric specifications.
21. Architects
Metric specifications and dimensions.
22. Restaurants
Menus will list metric units.
Recipes will use metric measures.
23. Insurance
Cost on land by square kilometers or hectares or square meters.
Costs on automobiles by kilometers driven per year.
24. Advertising
Measured in metric units.
25. Meteorologists
Many use metric already.
Metric distances.
26. Photography
Already metric.

As can be seen, occupations will be affected by the change to the metric system. The transition will be easier if we have a common frame of reference for the metric units. But measurements will be listed in both customary and metric measure during America's change to the metric system.

Appendix A Working with Powers of Ten

Many times it is necessary to work with very large and very small numbers. For example, the velocity of light is approximately 300 000 000 m/s or the mass of a proton is 0.000 000 000 000 000 000 000 001 67 g.

Large and small numbers can be expressed much more simply using powers-of-10 notation (often called scientific notation). This procedure is a method for keeping track of the decimal point when dealing with large numbers, small numbers, or when changing from one metric unit to another.

multiplication

The problem 10 × 10 × 10 can be written 10 × 10 × 10 = 10^3 = 1000. The 3 is called the exponent of 10 and indicates how many 10s are multiplied together. One hundred is the "second power" of 10 (10 × 10 = 10^2 = 100). Similarly 1000 is the "third power" of 10. See if you can see a relationship between the exponent and the number of zeros after the digit 1.

$$10^1 = 10$$
$$10^2 = 100$$
$$10^3 = 1000$$
$$10^4 = 10\ 000$$
$$10^7 = 10\ 000\ 000$$

Practice by writing the following numbers as powers of 10, using the appropriate exponent.

a. 1 000 000   b. 100 000 000   c. 100 000

Answers   a. 10^6 b. 10^8 c. 10^5

Change the following from exponential form to a simple numeral.

a. 10^4 b. 10^7 c. 10^{10}

Answers   a. 10 000   b. 10 000 000   c. 10 000 000 000

Products of powers of 10 can be found rapidly by noticing that $10^2 \times 10^3 = (10 \times 10) \times (10 \times 10 \times 10) = 10^5$.

$10^7 \times 10^3 = (10 \times 10 \times 10 \times 10 \times 10 \times 10 \times 10) \times (10 \times 10 \times 10) = 10^{10}$

For any natural number a and b

$$10^a \times 10^b = 10^{a+b}$$

therefore: $10^4 \times 10^5 = 10^{4+5} = 10^9$.

Practice by solving the following problems, using powers-of-10 notation.

a. $10^6 \times 10^4 =$ b. $10^1 \times 10^3 =$ c. $10^{23} \times 10\ 000 =$

Answers   a. 10^{10} b. 10^4 c. 10^{27}

division

If the numerator and denominator of a fraction are powers of 10, then there is a way of expressing the quotient as a power of 10.

$$\frac{10^5}{10^2} = \frac{10 \times 10 \times 10 \times 10 \times 10}{10 \times 10} = 10 \times 10 \times 10 = 10^3$$

$$\frac{10^6}{10^4} = \frac{10 \times 10 \times 10 \times 10 \times 10 \times 10}{10 \times 10 \times 10 \times 10} = 10 \times 10 = 10^2$$

For any natural numbers a and b

$$\frac{10^a}{10^b} = 10^{a-b}$$

therefore: $\dfrac{10^8}{10^5} = 10^{8-5} = 10^3$.

Practice by solving the following problems, using powers-of-10 notation.

$$a. \frac{10^4}{10} = \qquad b. \frac{10^7}{10^4} = \qquad c. \frac{10^8}{100} =$$

Answers a. 10^3 b. 10^3 c. 10^6

numbers between 1 and 0

The division rule allows the following relationships:

$$\frac{10^1}{10^1} = 10^0 = 1$$

$$\frac{1000}{10\ 000} = \frac{10^3}{10^4} = 10^{3-4} = 10^{-1} = 0.1$$

Set up a problem that would give these results:

$$10^0 = 1$$
$$10^{-1} = 0.1$$
$$10^{-2} = 0.01$$
$$10^{-4} = 0.0001$$
$$10^{-6} = 0.000\ 001$$

The number of zeros to the right of the decimal point is *one less than* the value of the exponent. The rules of multiplication and division of exponents remain the same.

Therefore 0.000 01 could be written 10^{-5}.

Practice by simplifying the following problems, using powers of **10** where appropriate.

a. $0.001 =$ b. $0.000\ 000\ 1 =$ c. $10^{-8} =$
d. $10^5 \times 10^{-4}$ e. $\dfrac{10^6}{10^{-4}}$ f. $10^{-2} \times 10^{-3} =$

Answers a. 10^{-3} b. 10^{-7} c. $0.000\ 000\ 01$
d. 10 or 10^1 e. 10^{10} f. 10^{-5}

For a problem such as 9.5 X 100 = 950, the result can be obtained by shifting the decimal point in 9.5 two digits to the *right.*

For a problem such as 950 ÷ 100 = 9.5, the result can be obtained by shifting the decimal point in 950 two digits to the *left.*

Practice by working the following problems:

a. 0.000 95 X 10 000 =   b. 9.5 X 0.001 =
c. 0.73 X 1000 =   d. 0.65 ÷ 1000 =
e. 4.23 ÷ 10 =   f. 283 ÷ 100 =

Answers a. 9.5   b. 0.009 5   c. 730
d. 0.000 65   e. 0.423   f. 2.83

In scientific notation most numbers are expressed as the product of two factors. For example: 3 300 000 would be written 3.3×10^6 and 0.000 42 would be written 4.2×10^{-4}. A number is written in scientific notation if it is in the form $a \times 10^y$. In this form a is an integer between 1 and 10, and y is an integer.

If it turns out that y is 0, the second factor is not written. If a is 1, it is not written.

Practice by writing the following numbers in scientific notation.

a. 52 600 =   b. 17 million =
c. 0.0039 =   d. 0.000 000 963 =

Answers a. 5.26×10^4 b. 1.7×10^7
c. 3.9×10^{-3} d. 9.63×10^{-7}

Since the metric system is a decimal system, the rules that apply to working with powers of 10 and scientific notation hold true.

In the metric system each unit is exactly 10 times larger than the next smaller unit. Also, any unit is a multiple of 10 larger than any smaller unit.

For example:

$$1 \text{ m} = 100 \text{ cm or } 10^2 \text{ cm}$$
$$1 \text{ km} = 1\ 000\ 000 \text{ mm or } 10^6 \text{ mm}$$

When going from a larger to a smaller unit multiply by the appropriate multiple of 10 (*move the decimal to the right*). Practice by making the following metric conversions. Write your answer as numerals and in scientific notation.

a. 43 m = _____ cm
b. 3 km = _____ m
c. 5.2 cm = _____ mm
d. 25 kg = _____ g
e. 22.4 liters = _____ ml
f. 4 km = _____ mm
g. 4 km = _____ cm

Answers   a. 4300 or 4.3×10^3
           b. 3000 or 3×10^3
           c. 52 or 5.2×10
           d. 25 000 or 2.5×10^4
           e. 22 400 or 2.24×10^4
           f. 4 000 000 or 4×10^6
           g. 400 000 or 4×10^5

Conversely, in the metric system each unit is exactly 10 times smaller than the size of the next larger unit. Any unit is a multiple of 10 smaller than any next larger unit.

For example:

$$1 \text{ m} = 1/1000 \text{ km or } 10^{-3} \text{ km}$$
$$1 \text{ mm} = 1/10 \text{ cm or } 10^{-1} \text{ cm}$$

When going from a smaller to a larger unit, move the decimal to the left.

Practice by making the following metric conversions. Write your answers as numerals and in scientific notation.

a. 43 cm = _____ m
b. 3 m = _____ km
c. 5.2 mm = _____ cm
d. 25 g = _____ kg

e. 22.4 ml = _____ liters
f. 4 mm = _____ km
g. 4 cm = _____ km

Answers   a. 0.43 or 4.3×10^{-1} b. 0.003 or 3×10^{-3}
     c. .52 or 5.2×10^{-1} d. 0.025 or 2.5×10^{-2}
     e. 0.0224 or 2.24×10^{-2} f. 0.000 004 or 4×10^{-6}
     g. 0.000 04 or 4×10^{-5}

Think carefully on these:

a. 1 m^2 = _____ cm^2
b. 1 cm^2 = _____ mm^2
c. 1 cm^2 = _____ m^2
d. 1 mm^2 = _____ m^2

Answers   a. 10 000 or 10^4 b. 100 or 10^2
     c. 0.0001 or 10^{-4} d. 0.000 001 or 10^{-6}

Appendix B A Diagnostic Test
of the Metric System

The following test can be used to determine areas of metric competencies. The numbers in parenthesis will indicate the expected number of problems necessary to indicate mastery.

decimals and powers of 10 (4 of 5 for mastery)

1. The number 8600 written in scientific notation would be
    a. 8.6×100
    b. 86×10^3
    c. 8.6×10^3
    d. 86×10^4
    e. none of these
2. If the number 1 000 000 000 were written in the form 10^n, n would have a value of
    a. 9
    b. 8
    c. 10
    d. 7
    e. none of these

3. The number 10^{-4} is also written
   a. 0.000 01
   b. 0.0001
   c. 10 000
   d. 10 × 4
   e. none of these
4. The number 1.5×10^{-6} could be written
   a. 15 000 000
   b. 0.000 015
   c. 1 500 000
   d. 0.000 001 5
   e. none of these
5. $500/10^3$ can be expressed as
   a. 5×10^{-1}
   b. 0.5
   c. 500/1000
   d. all of these
   e. none of these

metric linear measures (6 of 7 for mastery)

6. A meter is the same as
   a. 100 cm
   b. 1000 mm
   c. 1/1000 km
   d. all of the above
   e. none of the above
7. The basic unit of linear measure in the metric system is the
   a. gram
   b. cubit
   c. furlong
   d. meter
   e. none of these
8. To measure the thickness of a dime you would use _____ in the metric system.
   a. meters
   b. millimeters
   c. centimeters
   d. kilometers

e. none of these

9. A meter is nearest in length to
   a. a yard
   b. a foot
   c. a mile
   d. an inch
   e. none of these

10. A kilometer is a little longer than
    a. a yard
    b. a furlong
    c. one mile
    d. 0.5 mile
    e. none of these

11. A centimeter is a little less than
    a. a foot
    b. 0.5 inch
    c. a football field
    d. 0.5 mile
    e. none of these

12. Seven meters is *not* the same as
    a. 0.007 km
    b. 7000 mm
    c. 70 cm
    d. 70 dm
    e. none of these

metric mass measurements (4 of 5 for mastery)

13. A package of butter is measured in _____ in the metric system.
    a. kilograms or grams
    b. liters or milliliters
    c. meters or centimeters
    d. cubic centimeters
    e. none of these

14. A milligram is equal to
    a. 1000 g
    b. 1/1000 g
    c. 1/10 kg
    d. 10 dg

e. none of these
15. A kilogram is approximately equal to
    a. 0.5 pound
    b. 2 pounds
    c. 100 pounds
    d. 1000 pounds
    e. none of these
16. A nickel could serve as a good _____ mass piece.
    a. 1 ounce
    b. 1 g
    c. 5 g
    d. 5 grain
    e. none of these
17. 6500 g can be written as
    a. 6.5×10^2 g
    b. 6.5×10^{-3}
    c. 6.5 mg
    d. 6.5 kg
    e. none of these

metric volume and capacity measures (6 of 7 for mastery)

18. Which of the following is the metric unit for measuring liquid capacity?
    a. imperial gallon
    b. gram
    c. liter
    d. millimeter
    e. none of these
19. 1 l is equal to
    a. 1/1000 liter
    b. 100 ml
    c. 10 liters
    d. 1000 milliliters
    e. none of these
20. Complete the statement: 1 ml = _____ of water.
    a. 10 g or 10 cm^3

b. 1 g or 1 cm^3
c. 10 g or 10 cm^3
d. 1 g or 10 cm^3
e. none of these

21. Which of the following would be appropriate for measuring vanilla flavoring in a recipe?
   a. milligrams
   b. millimeters
   c. milliliters
   d. kilograms
   e. none of these

22. The volume of a brick would be measured in
   a. cubic kilometers
   b. cubic milliliters
   c. cubic centimeters
   d. cubic grams
   e. none of these

23. A cubic meter would be the same as
   a. 1 000 000 cm^3
   b. 10 000 000 mm^3
   c. 10^{-6} km^3
   d. all of the above
   e. none of the above

24. A liter container will hold slightly more than
   a. an imperial gallon
   b. a pint
   c. a quart
   d. a bushel
   e. none of these

metric area measure (three of four for mastery)

25. The area in Figure 31 is approximately
   a. 6 cm^2
   b. 30 square radians
   c. 2 mm^2
   d. 1/60 m^2
   e. none of these

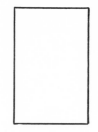

FIGURE 31.

26. The surface area of a book would likely be expressed in _____ in metric measure.
   a. square radians
   b. square milliliters
   c. square centimeters
   d. square meters
   e. none of these

27 Which of the following statements is *not* true?
   a. $1 \text{ cm}^2 = 100 \text{ mm}^2$
   b. $1 \text{ m}^2 = 1000 \text{ cm}^2$
   c. $1 \text{ m}^2 = 1\,000\,000 \text{ mm}^2$
   d. all are true
   e. none of these

28. A hectare is approximately
   a. 10 acres
   b. 100 m^2
   c. 2.5 acres
   d. $1\,000\,000 \text{ m}^2$
   e. none of these

metric temperature measure (4 of 4 for mastery)

29. The metric thermometer measures degrees
   a. Fahrenheit
   b. centigrade
   c. Celsius
   d. all of the above
   e. none of these

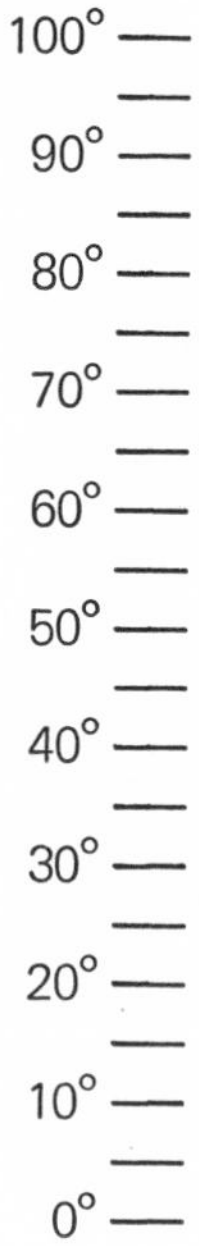

```
100° ——
       ——
 90° ——
       ——
 80° ——
       ——
 70° ——
       ——
 60° ——
       ——
 50° ——
       ——
 40° ——
       ——
 30° ——
       ——
 20° ——
       ——
 10° ——
       ——
  0° ——
```

FIGURE 32.

30-32. On the thermometer scale (Figure 32) mark
   a. a person's body temperature
   b. the boiling temperature of water
   c. the freezing temperature of water

Appendix C A Metric Mastery Test

1. The _____ is the basic unit for measuring length in the metric system.
2. The _____ is the basic unit for measuring mass in the metric system.
3. The _____ is the unit for measuring capacity in the metric system.
4. The _____ is used to measure temperature in the metric system.
5. To measure volume the answer would be expressed in _____ meters, _____ centimeters, etc.
6. To measure area the answer would be expressed in _____ meters, _____ centimeters, etc.
7. In the metric system each unit of length is _____ times the next smaller unit.
8. Write the symbol for;
   a. meter _____   e. liter _____
   b. kilometer _____   f. gram _____
   c. centimeter _____   g. Celsius _____
   d. millimeter _____   h. cubic centimeter _____
9. What does the symbol m^2 stand for? _____
10. Five cm^3 is the same as _____ mm^3.

11. Write in the unit you would use to measure the following:
    a. container of milk _____
    b. a beef roast _____
    c. a length of rope _____
    d. a truckload of sugar beets _____
    e. a zipper _____
    f. a lawn _____
    g. room temperature _____
    h. electricity used _____

12. Match the approximate mass with the object.
    _____ a. 50 ml of water   1. 2 kg
    _____ b. a nickel   2. 80 kg
    _____ c. a grown man   3. 50 g
    _____ d. a small bag of sugar   4. 5 g
    _____ e. a large car   5. 2 t

13. Match the event with the most appropriate temperature.
    _____ a. water freezes   1. $100°C$
    _____ b. a comfortable room   2. $37°C$
    _____ c. water boils   3. $0°C$
    _____ d. a cake bakes   4. $21°C$
    _____ e. normal body temperature   5. $160°C$

14. Complete the following:
    a. 1.6 m = _____ km = _____ cm
    b. 0.4 cm = _____ m = _____ mm
    c. 5 cm^2 = _____ mm^2 = _____ m^2
    d. 260 g = _____ kg
    e. 22.4 l = _____ ml
    f. 1 m^3 = _____ cm^3 = _____ mm^3
    g. 2.2 kg = _____ g = _____ mg

15. List an approximate customary system relationship for each of the following:
    a. A meter is _____
    b. A kilogram is _____
    c. A liter is _____
    d. A centimeter is _____

16. Five g of water is about _____ ml = _____ cm^3

1. meter
2. kilogram
3. liter
4. Celsius or Kelvin thermometer
5. cubic, cubic
6. square, square
7. 10
8. a. m
   b. km
   c. cm
   d. mm
   e. l
   f. g
   g. C
   h. cm^3
9. square meter
10. 5000
11. a. l
    b. kg
    c. m
    d. metric ton
    e. cm
    f. square meters

g. degree Celsius
h. kilowatt

12. $\underline{3}$ a.
    $\underline{4}$ b.
    $\underline{2}$ c.
    $\underline{1}$ d.
    $\underline{5}$ e.

13. $\underline{3}$ a.
    $\underline{4}$ b.
    $\underline{1}$ c.
    $\underline{5}$ d.
    $\underline{2}$ e.

14. a. 0.001 6, 160
    b. 0.004, 4
    c. 500, 0.000 5
    d. 0.26
    e. 22 400
    f. 10^6, 10^9
    g. 2200, 2 200 000

15. a. a little longer than a yard
    b. a little heavier than 2 pounds
    c. a little more than a quart
    d. a little less than 0.5 inch

16. 5, 5

Appendix D Metric Bibliography

"Adult Education and the Metric System," Richard W. Cartright. *Adult Leadership,* 20:190, November 1971.

Aids for the Age of Metrics. Oak Lawn, Illinois: Ideal School Supply Co.

American National Standard Metric Practice Guide: Z210.1-1973 (ASTM E380-72): American National Standards Institute.

Amusements in Developing Metric Skills, Alice A. Clark and Carol H. Leitch. Troy, Michigan: Midwest Publications, 1973.

"An American Dilemma: Measuring Up in the Future," Gerald W. Brown. *School Science and Mathematics,* 71:435-436, May 1971.

American Metric Journal (bimonthly), $35/yr., AMJ Publ. Co., Drawer L., Tarzana, California 91356.

Antitrust Implications of Metric Conversion, American National Standards Institute.

"Apparatus Review," D.J. Maxwell. *Mathematics in School,* 1:33, November 1971.

"Are Inches, Pints, Pounds on the Way Out in the U.S.?" *U.S. News and World Report,* 71:73-74, 1972.

"Are You Ready for the Meter?" John Teresko. *Industry Week,* May 1973.

ASTM Standard Metric Practice Guide. Philadelphia: American Society for Testing and Materials.

"Bilingualism in Measurement: The Coming of the Metric System," G.W. Bright. *Arithmetic Teacher,* 20:397-399, May 1973.

"Brief History and Use of the English and Metric System of Measurement," *Science Teacher,* 36:39-40, May 1969.

Brief History of Measurement Systems, National Bureau of Standards Special Publication (0303-01073).

Brief History of Measurement Systems With a Chart of the Modernized Metric System. National Bureau of Standards (U.S.) Spec. Publ. 304A, 3 pages (1972), 30 cents (SD Catalog No. C13.10:304A).

"The Case for Metric Units," Tommie A. West. *School Science and Mathematics,* 72:600-02, October 1972.

"Changing to the Metric System: An Idea Whose Time Has Come," R.W. Joly. *NASSP Bulletin,* 56:47-59, November 1972.

Characteristics of the Metric System. Chicago, Illinois: Society for Visual Education.

Classroom Metric Lines. Duluth, Minnesota: The Instructor Publications, Inc.

Color Me Metric. Carole Gould. San Jose: A.R. Davis and Company, 1973.

"Coming of the Metric System," Frank Kendig. *Science,* November 25, 1972.

Commercial Weights and Measures (An Interim Report of the U.S. Metric Study), S.L. Hatos. Nat. Bur. Stand. (U.S.) Spec. Publ. 345-3, 109 pages (July 1971), $1.00 (S.D. Catalog No. C13.10:345-3).

The Consumer (An Interim Report of the U.S. Metric Study), B.D. Rothrock. Nat. Bur. Stand. (U.S.) Spec. Publ. 345-7, 146 pages (July 1971), $1.25 (SD Catalog No. C13.10:345-7).

Converting to Metric. Santa Monica, California: BFA Educational Media.

Cuisenaire Rod Math Kits. New Rochelle, New York: Cuisenaire Co. of America.

"Decimalization and Metrication in the Gas Industry," John Oliver. *Industrial Training International,* 6:358-59, December 1971.

Decimeter (game), Paul Ploutz. Athens, Ohio: The Lawhead Press.

"Decimals Before Fractions," Jack Berryman. *Mathematics in School,* 1:18-20, July 1972.

Department of Defense (An Interim Report of the U.S. Metric Study), L.E. Barbrow, coordinator. Nat. Bur. Stand. (U.S.) Spec. Publ. 345-9, 125 pages (July 1971), $1.25 (SD Catalog No. C13.10:345-9).

"A Descriptive Analysis of the Teaching of the Metric System in the Secondary Schools," Mary Oellerich Murphy and Maxine A. Polzin. *Science Education,* 53:89-94, February 1969.

Discover Why Metrics. Beloit Tool Company. Roscoe, Illinois: Swani Publishing Company, 1972.

Discovering Metric Measure (intermediate workbook), Gary G. Bitter, Jerald L. Mikesell, and Kathryn Maurdeff. New York: Webster/McGraw Hill, 1975.

Education (An Interim Report of the U.S. Metric Study), B.D. Robinson. Nat. Bur. Stand. (U.S.) Special Publ. 345-6, 209 pages (July 1971), $1.25 (SD Catalog No. C13.10:345-6).

"Education and Training in SI Units," L.F. Sokol. *American Metric Journal,* January and February 1974.

"Education in Decimal Currency and the Metric System; A Report on Great Britain's Progress in the Elementary Schools," Evan E. McFee. *School Science and Mathematics* 69:644-6, October 1969.

"The Effect of Metrication on the Consumer," J.V. Odom. *Business Education Forum,* December 1973, p. 10.

Electrical Engineering Units and Constants (pocket card), Nat. Bur. Stand. (U.S.) Spec. Publ. 368 (1972), 10 cents each or $6.25 per 100 (SD Catalog No. C13.10:368).

Engineering Standards (An Interim Report of the U. S. Metric Study), R.D. Stishler. Nat. Bur. Stand. (U.S.) Spec. Publ. 345-11, 257 pages (July 1971), $2.00 (SD Catalog No. C13.10:345-11).

"Experience, Key to Metric Unit Conversion," Malcolm D. Swan. *Science Teacher,* 37:69-70, November 1970.

"Experiences for Metric Missionaries," L. Viets. *Arithmetic Teacher,* 20:269-73, April 1973.

Experience in the Metric System, H.N. Johnson and M.J. Robeson. Minneapolis: Paul S. Amidon and Associates, 1974.

Exploring Metric Measure—Primary Teacher's Source Book, John H. Bates. New York: McGraw-Hill, 1972.

Exploring the Metric System, Kemp and Richards. River Forest, Illinois: Laidlaw Brothers, 1973.

Exploring with Metrics, Gary G. Bitter and Tom Metos. New York: Julian Messner, 1975.

Federal Government: Civilian Agencies (An Interim Report of the U.S. Metric Study), R.E. Clark and J.M. Taschler. Nat. Bur. Stand. (U.S.) Spec. Publ. 345-2, 324 pages (July 1971), $2.25 (SD Catalog No. C13.10345-2).

Food Preparation—Recipes and Their Conversion, Fern E. Hunt. American Home Economics Association, 1973.

"The Foot, Quart, and Pound Collide with Computation, U.S. Goes Metric," Paul Shoecraft. *School Science and Mathematics,* 64:67-68, January 1974.

For Good Measure: Compare Metric and Customary Units (pocket ruler). Nat. Bur. Stan. (U.S.) Spec. Publ. 376 (1972), 10 cents (SD Catalog No. C13.10:376).

"Ford's New Metric Engine," *Quality Management and Engineering,* September 1973, pp. 14-18.

"From Finger Counting to the Metric System," Frost, Douglas V. *Applied Optics,* 5:1257-28, August 1966.

Fun with Metric Measurement, Betty Long and Carol Witte. Manhattan Beach, California: Teachers, 1973.

"Get Ready for the Metric System," Richard D. Bowles. *Instructor,* 81:69-70, December 1971.

Go Metric. (Bumper Stickers). Oak Lawn, Illinois: Ideal School Supply Co.

Go Metric. (Lapel Pin). Oak Lawn, Illinois: Ideal School Supply Co.

Going Metric. (Cassette Learning Package) Wilkes-Barre, Pennsylvania: Media Materials.

"Going Metric All the Way," *ASTM Standardization News,* 1:19-22, July 1973.

"Going Metric—Going Decimal," R.W. Shaw. *Mathematics in Schools,* 1:23-24, November 1971.

"Going Metric Makes the Figuring Easier in Your Shop," Marshall Lincoln. *Popular Science,* November 1973, pp. 127-29.

Guide to Impact of Metric Use on Standards Development in Companies, Trade Associations, Technical and Professional Societies: American National Standards Institute.

Happy Metric Books. Ft. Collins, Colorado: Scott Resources, Inc., 1975.

"Here Comes the New 'Yardstick' in Your Life," Edward Edelson. *Popular Science,* pp. 80-83, 150; November 1973.

A History and Overview of Metrication and Its Impact on Education, Jeffrey Odom. National Bureau of Standards, 1972.

A History of the Metric Controversy in the United States (An Interim Report of the U.S. Metric Study), C.F. Treat. Nat. Bur. Stand. (U.S.) Spec. Publ. 345-10, 306 pages (Aug. 1971), $2.25 (SD Catalog No. C13.10:345-10).

History of the Metric System. Chicago, Illinois: Society for Visual Education

How Much Can it Hold? Jamaica, New York: Eye Gate House.

How Much Does it Weigh? Jamaica, New York: Eye Gate House.

How Tall, How Far, and How Fast? Jamaica, New York: Eye Gate House.

How to Convert the Metric System into the U.S. System—Vice Versa. Warwick, New York: Real-T-Facs.

"How to Teach High School Students Any Metric Units They Need to Know," *Education,* 71:491-492, Stanford L. Kunkle, April 1951.

How You Could Possibly Live with the Metric System, John E. Robinson. Orange, California: Human Design Advocates, 1972.

"Ideas," George Immerzeel. *Arithmetic Teacher,* 20:280-87, April 1973.

"Ideas for Teachers," George Immerzeel and Don Wiederanders. *Arithmetic Teacher,* 19:362-63, May 1972.

"IFI Pushes Cooperative Development of Metric Standards," *Product Engineering,* pp. 13-14, October 1973.

I'm O.K.—You're O.K., Let's Go Metric, Donald Buckeye. Troy, Michigan: Midwest Publications, 1973.

"Inching our Way Toward the Metric System," G. Vervoort. *Mathematics Teacher.* 66:297-302, April 1973.

Information on the Metric System and Related Fields, Ernst Lange. George C. Marshall Space Flight Center, Marshall Space Flight Center, Alabama, 6th Edition, 1974.

Information on the Metric System and Related Fields. Boulder, Colorado: U.S. Metric Association.

Insight Into Metric Measurement. San Jose, California: Parks and Math Company, 1974.

The International (SI) Metric System and How it Works, Robert A. Hopkins. Tarzana, California: Polymetric Services, Inc., 1975.

International Standards (An Interim Report of the U.S. Metric Study), R.D. Huntoon *et al.* Nat. Bur. Stand. (U.S.) Spec. Publ. 345-1, 145 pages (Dec. 1970), $1.25 (SD Catalog No. C13.10:345-b).

"The International Systems of Units (SI)," Chester H. Page. *Physics Teacher,* 9:379-81, October 1971.

The International System of Units (SI). Nat. Bur. Stand. (U.S.), Spec. Publ. 330, 42 pages (1972), 30 cents (SD Catalog No. C13.10:330/2).

International Trade (An Interim Report of the U.S. Metric Study), G.F. Gordon. Nat. Bur. Stand. (U.S.) Spec. Publ. 345-8, 181 pages (Aug. 1971), $1.50 (SD Catalog No. C13.10:345-8).

Introducing the Metric System with Activities, D.A. Buckeye. Troy, Michigan: Midwest Publications, 1972.

Introducing the Metric System. Santa Monica, California: BFA Educational Media.

"The Introduction of Metric By the Use of Comics," Gunter Picket. *Educational Studies in Mathematics,* 4:31-47, June 1971.

Introduction to the Metric System. New Rochelle, New York: Pathescope Educational Films, Inc.

Investigating Metric Measure (junior high workbook), Gary G. Bitter, Jerald L. Mikesell, and Kathryn G. Maurdeff. New York: McGraw-Hill, 1975.

"Is Metric a Measure of Pain?" Orville Nelson. *Industrial Arts and Vocational Education,* 61:22-24, 32, February 1972.

ISO Standard 1000, SI Units and Recommendations for the Use of Their Multiples and of Certain Other Units. American National Standards Institute.

Know the Essentials of Metric Measurement, Book I. Wilkinsburg, Pennsylvania: Hayes School Publishing.

Learn Metrics by Doing Metrics. Stockton, California: Willow House Publishers, 1974.

Length, Width, Height, and Speed. Jamaica, New York: Eye Gate House.

Let's Go Metric, Frank Donovan. New York: Weybright and Talley, 1973.

Let's Play Games in Metrics, George L. Henderson and Lowell D. Glunn. Skokie, Illinois: National Textbook Company, 1974.

"Let's Start Measuring Up to the Metric Scale," C. Grieder. *Nations Schools,* 88:26, November 1971.

"Let's Start Metrics Now," Oliver Oberlander. *School Shop,* 31:26-27, June 1972.

"Let's Teach the Metric System Through Its Use," L.C. Oberline. *Arithmetic Teacher,* 14:376, May 1967.

Liter Volume Set. Palo Alto, California: Creative Publications, Inc.

Making Metric Maneuvers, Mary Richardson Miller and Toni Cresswell Richardson. Hayward, California: Activity Resources Company, 1974.

The Manufacturing Industry (An Interim Report of the U.S. Metric Study), L.E. Barbrow, coordinator. Nat. Bur. Stand. (U.S.) Spec. Publ. 345-4, 165 pages (July 1971), $1.25 (SD Catalog No. C13.10:345-4).

MathMetrics, Charles Geer and John Geer. Burlingame, California: Modern Math Materials, 1975.

Mass and Weight. Jamaica, New York: Eye Gate House.

Math and Metric Dictionary. San Jose, California: Parks and Math Company, 1974.

The Meaning of Metric. Santa Monica, California: BFA Educational Media.

"Meaningful Metric," Ed Strong. *School Science and Mathematics,* 64:421-22, May 1964.

Measure for Measure: A Guide to Metrication for Workshop Crafts and Technical Studies. Schools Council (London). New York: Citation Press, 1970.

Measure Metric (Elementary Workbooks), Lola J. May and Donna C. Jacobs. New York: Harcourt, Brace and Javonovich, 1973.

Measurement and Metric System Science Packet. Washington: National Science Teacher's Association.

Measurement for Everything. Wichita, Kansas: Library Filmstrip Center.

Measurement Skills. Chicago: Encyclopedia Britannica Educational Corporation.

"Measurement Standards, Physical Constants, and Science Teaching," E. Ambler. *Science Teacher,* 38:63-71, November 1971.

Measuring A 'La Metric. San Jose, California: Parks and Math Company, 1974.

Measuring a Metric Approach. Tarrytown, N.Y.: Schloat Productions, 1974.

"Measuring Metrically Pleasurably," Gary G. Bitter and Thomas H. Metos. *Educating Children: Early and Middle Years,* 19:11-14, Summer 1974.

"Measuring Up to Metric," Lawrence M. Kushner. *Highlights,* pp. 14-15, Dec., 1972.

Measure With Metric. New York: Thomas Y. Crowell, 1975.

Merry Metric Activities. Fresno, California: Teaching Associates, 1973.

Meter- Liter-Gram. Warwick, New York: Real-T-Facs.

Meter Means Measure, Carl S. Hirsch. New York: Viking Press, 1972.

"The Meter Stick," Susan G. Sheffield. *Science and Children,* pp. 22-24, March 1973.

"Meters Come to Malibu," *Evening Outlook,* March 14, 1974.

Meters, Liters, and Grams. School Council, (London). New York: Citation Press, 1970.

Meters, Liters, and Grams, Audrey V. Buffington. New York: Random House, 1973.

Metric Activities (Workbook), Lee Hutton. Toronto, Canada: Holt, Rinehart and Winston, 1973.

"Metric A Go-Go," *Science,* 175:1346, March 1972.

A Metric America: A Decision Whose Time Has Come, Daniel V. DeSimone. Washington: Government Printing Office, 1971.

Metric Association Newsletter, (quarterly) $3/year, U.S. Metric Association, Inc., Sugarloaf Start Route, Boulder, Colorado 80302.

"Metric Bill Finally Voted," *Science News,* 102:132, 1972.

The Metric Center (kit). Palo Alto, California: Enrich.

"The Metric Change Over," *National Bureau of Standards Technical News Bulletin,* 57:103-106, May 1973.

Metric Cluster. S-APA, Lexington, Massachusetts: Ginn and Co.

Metric Conversion and You. Metric Conversion Board, (Australia), 1970.

Metric Conversion Card (pocket card). Nat. Bur. Stand. (U.S.) Spec. Publ. 365 (1972) 20 cents domestic postpaid, or 10 cents GPO Bookstore (SD Catalog No. C13.10:365/2).

"Metric Conversion: The Training Colossus of the Seventies," Joseph L. Pokorney. *Training and Development Journal,* June 1973.

The Metric Fallacy, F.A. Halsey. Washington: The American Institute of Weights and Measures, 1920.

Metric Fun, Betty Long and Carol Witte. Manhattan Beach, California: Teachers, 1974.

Metric Guide Bulletin, Neenah, Wisconsin, J.J. Keller and Associates.

Metric Handbook for Hospitals. Oak Lawn, Illinois: Ideal School Supply Co.

Metric In Progress. Wichita, Kansas: Library Filmstrip Center.

Metric is Coming. National Science Teacher's Association, Washington, D.C., 1973.

Metric Ease. Palo Alto, California: Creative Publications, 1975.

"Metric is Here: So Let's Get on With It," R. Fisher. *Arithmetic Teacher,* 20:400-402, May 1973.

Metric Linear Measurement Set. Palo Alto, California: Creative Publications, Inc.

Metric Measurement. Kankakee, Illinois: Imperial International Learning Corporation.

Metric Measurement, Cecil R. Trueblood. Dansville, New York: The Instructor Publications, 1973.

Metric Measurement. Minneapolis: Minneapolis Public Schools, 1973.

Metric Measurement. Chicago: Educational Teacher Aids, 1974.

The Metric Mice Measure, Marty Hiatt and Linda Harvey. Carson, California: Educational Insights, 1974.

Metric Minder Kit. Barrington, New Jersey: Edmund Scientific.

Metric Modules—Preservice and Inservice Programs. Oviedo, Florida: Kent Education Service.

Metric *News* (bimonthly), $5/year, Swani Publishing Co., Box 248, Roscoe, Illinois 61073.

"Metric Not If, But How," NCTM Metric Implementation Committee, *The Arithmetic Teacher,* pp. 366-69, May 1974.

Metric Place Value Chart. Oak Lawn, Illinois: Ideal School Supply Co.

"Metric Plan Bill is Killed in House by 240-153 Vote," *Wall Street Journal,* May 8, 1974.

Metric Power, Richard Deming. New York: Thomas Nelson Co., 1973.

Metric *Reporter* (biweekly), American National Metric Council, $25/year, 1625 Massachusetts Ave., NW, Washington, D.C. 20036.

Metric Sampler. Tuxedo, New York: Union Carbide Educational Aids Department.

Metric Songs. Commack, N.Y.: Great Ideas, 1974.

"Metric Students and You!" *Instructor,* pp. 60-65, October 1973.

Metric Supplement to Science and Mathematics (Workbook), Fred J. Helgren. Oak Lawn, Illinois: Ideal School Supply, Inc., 1973.

"The Metric System," Franklyn M. Branley. *Grade Teacher,* 76:49, 100, Dec. 1959.

The Metric System. Dansville, New York: The Instructor Publications, 1973.

The Metric System (junior high workbook). Menlo Park, California: Addison Wesley, 1974.

Metric System Ahead. NEA Research Bulletin, 49:109-112, December 1971.

"The Metric System and the Sew Set," Mildred C. Ryan (speech). Washington, D.C., American Home Economics Association, 1973.

The Metric System: A Programmed Approach, Paul F. Ploutz. Columbus, Ohio: Charles E. Merrill, 1972.

"Metric System Authorized," *California Business and Professions Code—* Section 15: 203.

Metric System Guide. Neenah, Wisconsin: J.J. Keller and Associates, 1973.

The Metric System, How to Use It. New York: McGraw-Hill Films.

"The Metric System in Elementary Grades," Fred J. Helgren. *The Arithmetic Teacher,* 14:349-53, May 1973.

"Metric System in Grade Six," D.R. Bowles. *The Arithmetic Teacher,* 11:36-38, January 1964. Met

"The Metric System in Junior High School," Isabelle P. Rucker. *Mathematics Teacher,* pp. 621-23, December 1958.

"The Metric System: Its Relation to Mathematics and Industry," Wilmer Souder. *Mathematics Teacher,* pp. 25-35, September 1920.

"The Metric System—Let's Emphasize Its Use in Mathematics," F.D. Alexander. *The Arithmetic Teacher,* 20:395-96, May 1973.

Metric System Library. Neenah, Wisconsin: J.J. Keller and Associates.

The Metric System of Weights and Measures: Twentieth Yearbook. Compiled by The Committee on the Metric System, J.T. Johnson, Chairman. New York: Columbia University, 1948.

"The Metric System: Past, Present, and Future?" A.E. Hallerberg. *The Arithmetic Teacher,* 20:247-55, April 1973.

The Metric System Simplified, Gerald W. Kelly. New York: Sterling Publishing Company, 1973.

The Metric System, Why Have It? New York: McGraw-Hill Films.

The Metric System with Activities (workbook), Donald A. Buckeye. Troy, Michigan: Midwest Publications Company, 1973.

Metrication (game). Orlando, Florida: Metrix Corporation.

"Metrication Activities in Education," *Business Education Forum,* December 1973, pp. 14-16.

"Metrication and Social Change," H.G. Forst. *Adult Education,* (London), 43:238-42, November 1970.

Metrication at Work, Training Manual, Local Government Training Board, (London), 1971.

Metrication for the Housewife, South African Bureau of Standards, 1974.

"Metrication in Britain," Elizabeth Williams. *Arithmetic Teacher,* pp. 261-64, April 1973.

"Metrication in the School Curriculum," E. Briggs. *Trends in Education,* 26:35-40, April 1972.

"Metrication: New Dimensions for Practically Everything," Lee Edson. *American Education,* 8:10-14, April 1972.

Metrication of America. New York: Westinghouse Learning Corp.

"Metrication—Our Responsibility?" Klaus E. Kroner. *Journal of Engineering Education,* 63:53-54, October 1972.

"Metrication Urged by N.S.T.A. Committee," *Science Teacher,* 38:6-7, Jan. 1971.

Metrics: An Introduction to the Metric System, Arthur E. Ring. Campbell, California: The Classroom Service Company, 1973.

"Metrics are Coming," B. Crane. *Grade Teacher,* 88:88-89, February 1971.

"The Metrics are Coming," *Changing Times,* pp. 33-34, May 1974.

Metrics Made Easy. Grand Rapids, Michigan: Instructional Fair.

"Metrics: Your Schools Will Be Teaching It and You'll be Living It—Very,

Very, Very Soon," P.G. Jones. *American School Board Journal,* 1670:21-25, July 1973.

Metric Units in Primary Schools, M.J. Lighthill *et al.* Royal Society, (London), April 1970.

Metric Units in the Science Laboratory. New Rochelle, New York: Pathescope Educational Films.

Metric Units of Area and Volume. New Rochelle, New York: Pathescope Educational Films.

Metric Units of Capacity. New Rochelle, New York: Pathescope Educational Films.

Metric Units of Length. New Rochelle, New York: Pathescope Educational Films.

Metric Units of Measure. Oak Lawn, Illinois: Ideal School Supply Co.

Metric Units of Measure. Metric Association of America, 1972.

Metric Weights and Measures. Wichita, Kansas: Library Filmstrip Center.

A Metric Workbook for Teachers of Consumer and Homemaking Education, C. Bielefeld. Santa Ana, California: Orange County Department of Education, 1973.

Metrikit. Columbus, Ohio: Charles E. Merrill Publishing Co.

Metrikit. Villa Park, Illinois: Larry Harkness Company.

Metrikit— Mini-Course. Villa Park, Illinois: Larry Harkness Co.

Mini-Metric Lab. St. Charles, Illinois: Aero Educational Products Ltd.

"A Miss Is as Good as a Kilometer?" *Interchange,* Transportation Department of California, Winter 1974.

Modern Math and Metric Skills-Building Workbook. San Jose, California: Parks and Math Co., 1974.

Modernized Metric System (0303-01072). Pueblo, Colorado. Public Documents Distribution Center.

"The Move to Metric, Viewing the Problems," R.G. Green. *Automation,* pp. 70-76, September 1973.

Moving Toward Metric. Chicago: J.C. Penney Co. Inc., 1974.

Mr. Windbag in Metricland. Oak Brook, Illinois: Educational Products, 1973.

My Metric Measurement Manual, Fritz Willert. Two Rivers, Wisconsin: Pauper Press, 1973.

NBS U.S. Metric Study Interim Report: A History of the Metric Controversy in the U.S. Washington: Government Printing Office, 1971.

"NBS Urges 10-Year Metric Conversion Plan," Constance Holden. *Science,* 173: 613, August 1971.

"New Dimensions for Practically Everything: Metrication," L. Edson. *American Education,* 8:10-14, April 1972.

"The New Push for the Metric System: Will You give Up Pounds, Feet, and Inches?" A.P. Armagnac. *Popular Science,* 204:54-57, 1969.

Nonmanufacturing Businesses (An Interim Report of the U.S. Metric Study), E.D. Bunten and J.R. Cornog. Nat. Bur. Stand. (U.S.) Spec. Publ. 345-5, 192 pages (Aug. 1971), $1.50 (SD Catalog No. C13.10:345-5).

Nova Scotia Measures Up: Background to Metrication. Truro, Nova Scotia: Continuing Education Division, Nova Scotia Teacher's College.

The Open Forum: The Current Status of Metric Conversion, J.V. Odom. Metric Information Office.

Orientation for Company Metric Studies (Mechanical Products Industry), 2nd Edition: American National Standards Institute, March 1, 1970.

"Our Clock Says '13' . . . And It's Time to Talk Metric," Don Harold Allen. *Kappa Delta Pi Record,* pp. 4-5, October 1973.

"Overcoming the Resistance to the Metric System," H. Balleu. *School Science and Mathematics,* 73:217-23, March 1973.

Practice in the Metric System. Dansville, New York: The Instructor Publications, Inc. 1973.

Prepare Now For A Metric Future, Frank Donovan, New York: Weybright and Tally, 1970.

"Preparing Now for a Metric Cubit," S.H. Sewell. *Man/Society/Technology,* 10:177-80, January 1973.

"Reading the Meter," *Newsweek,* pp. 59-61, February 18, 1974.

"The Relative Merits of Two Methodologies of Teaching the Metric System to Seventh-Grade Science Students," Evan E. McFee. Unpublished doctoral dissertation, Indiana University, Bloomington, Indiana, 1967.

"A Review of Research on the Teaching of the Metric System," Mary Oellerich Murphy and Maxine A. Polzin. *Journal of Educational Research,* 152:267-70. February 1969.

"A Right Now Project: How to Get Ready to Go Metric in Your School District," *American School Board Journal,* 160:26, July 1973.

"Schools are Going Metric," Fred J. Helgren. *The Arithmetic Teacher,* pp. 265-67 April 1973.

"SI In Engineering Education," Cornelius Wandmacker. *Journal of Engineering Education,* 61:827-30, April 1971.

SI: Length and Area. Wichita, Kansas: Library Filmstrip Center.

SI: Mass and Volume. Wichita, Kansas: Library Filmstrip Center.

SI Metric: Reference Manual. Armonk, New York: International Business Machines, 1973.

SI Metric: Style Manual. White Plains, New York: International Business Machines.

SI Metric: The International System of Units. White Plains, New York: International Business Machines.

SI Metric Units, An Introduction, H.F.R. Adams. Ontario, Canada: McGraw-Hill, Fyerson Ltd.

SI: Temperature. Wichita, Kansas: Library Filmstrip Center.

Simple Simon's Troubles. Wichita, Kansas: Library Filmstrip Center.

Sir Mortimer's New Measures. Wichita, Kansas: Library Filmstrip Center.

Sir Mortimer's Return. Wichita, Kansas: Library Filmstrip Center.

"Start Now to Think Metric," M. Warning. *Journal of Home Economics,* 64:18-21, December 1972.

The Story of Standards, J. Perry. New York: Funk and Wagnalls Company, 1955.

"A Study of the Case for Measurement in Elementary School Mathematics," Lloyd Scott. *School Science and Mathematics.* 66: 714-22, November 1966.

"A Study of Weights and Measures," Gertrude Cushing Yorke. *The Mathematics Teacher,* pp. 125-128, March 1944.

"A Study of Weights and Measures (Another View,)" J.T. Johnson. *The Mathematics Teacher,* pp. 219-21, May 1944.

Survey of Modernized Metric. Wichita, Kansas: Library Filmstrip Center.

Teaching the Metric System. Henry Boyd, Chicago: Weber Costello, 1974.

"Teaching the Metric System " T.W. Jeffries. *The Science Teacher,* 28:53, Feb. 1961.

"Teaching the Metric System as Part of Compulsory Conversion in the U.S." V.J. Hawkins. *The Arithmetic Teacher,* 20:390-94, May 1973.

Ten Times Ten is Simple. Wichita, Kansas: Library Filmstrip Center.

Testimony of Nationally Representative Groups (An Interim Report of the U.S. Metric Study), J.V. Odom, editor. Nat. Bur. Stand. (U.S.) Spec. Publ. 345-12, 174 pages (July 1971), $1.50 (SD Catalog No. C13.10:345-12).

"Think Metric," *NEA Research Bulletin,* 49:30-42, May 1971.

Think Metric. Oak Lawn, Illinois: Educational Products, Inc.

Think Metric! Franklyn M. Branley. New York: Thomas Y. Crowell and Company, 1972.

Think Metric, F.O. Armbruster. San Francisco: Troubador Press, 1973.

"Think Metric to Meet the Challenge," June Patterson. *Co-Ed,* September 1972.

Thinking Metric, Thomas Gilbert and Marilyn Gilbert. New York: John Wiley and Sons, 1973.

"Three Studies on the Effect of the Compulsory Metric Usage," Gertrude Cushing Yorke, *Journal of Educational Research,* XXXVII, pp. 343-352, January 1944.

"Three Studies on the Effect of the Compulsory Metric Usage: Another View," J.T. Johnson. *Journal of Educational Research,* XXXVII, pp. 575-588, April 1944.

Toward A Metric America. NBS list of publications. Washington: Government Printing Office, 1973.

"Training for Metrication," Lincoln Ralpho. *Industrial International,* 6:360-61, December 1973.

The Trouble with Measurement. Wichita, Kansas: Library Filmstrips Center.

"Two Landmark Science Bills pass Senate," *Science News,* 102:132, August 1972.

"Two-Pan Weighings," Chris Burditt. *Two-Year College Mathematics Journal,* 3:80-81, February 1972.

Understanding the Metric System. Chicago: Encyclopedia Britannica Educational Corporation.

Understanding the Metric System. Wethersfield, Connecticut: Janus Association.

Understanding the Metric System: A Programmed Approach, David Monroe Miller. Boston, Massachusetts: Allyn and Bacon, Inc., 1973.

Units of Weight and Measure—International (Metric) and U.S. Customary Definitions and Tables of Equivalents. Nat. Bur. Stand. (U.S.) Misc. Publ. 286, 251 pages (1967), $2.25 (SD Catalog No. C13.10:286).

"The U.S. Metric Study," Lewis M. Branscomb. *Science Teacher,* 38:58-62, November 1971.

U.S. Metric Study Interim Report: Nonmanufacturing Businesses. Washington: Government Printing Office, 1971.

U.S. Metric Study Interim Report: Testimony of Nationally Representative Groups. Washington: Government Printing Office, 1971.

"U.S. Moving, Inch by 2.54 cm to Metric System," Frank Donovan. *The Mainliner* United Airlines Magazine, pp. 24-27, July 1973.

USA Goes Metric (metric conversion kit). Barrington, New Jersey: Edmund Scientific Co.

The Use of SI Units. British Standards Institution, (London), January 1969.

"Use of the Metric System in Microbiology," Eugene Weinberg. *The American Biology Teacher,* 22:340-342, June 1960.

Using the Metric System, Kemp and Richards. River Forest, Illinois: Laidlaw Brothers, 1973.

VNR Metric Handbook, Leslie Fairweather and Jan A. Sliwa. New York: Van Nostrand Reinhold Co., 1969.

Volume and Capacity. Jamaica, New York: Eye Gate House.

Wall Chart of the Modernized Metric System (color). Nat. Bur. Stand. (U.S.) Spec. Publ. 304, 1 page (1972), 55 cents (SD Catalog No. C13.10: 304).

What About Metric? Louis E. Barbrow. Nat. Bur. Stand. (U.S.) Con. Infor. Set. 7 (1973), 80 cents (Stock No. 0303-01101).

"What It will Cost to Go Metric," *American School Board Journal,* 160: 25-26, July 1973.

"Where Industry Stands On Metrication," *Metal Progress,* pp. 67-70, August 1973.

"Why Convert to the Metric System Now?" L.M. Rhodes, *The Kiwanis Magazine,* pp. 19-23, March 1974.

Why Do We Measure? Jamaica, New York: Eye Gate House.

Why Go Metric? Chicago, Illinois: Society for Visual Education.

Why Metric? Jamaica, New York: Eye Gate House.

"Will the U.S. Go Metric?" Lewis M. Branscomb. *1973 Britannica Yearbook of Science and the Future.*

"You and the Metric System" Eileen D. Wray. *The Arithmetic Teacher,* pp. 576-80, December 1964.

Appendix E Metric Suppliers

Aakron Rule Corporation
59 Hoag Avenue
Akron, NY 14001

A. Balla and Company
P.O. Box 24200
Ft. Lauderdale, FL 33307

Abbey Books
Metrick Media Book Publishers
P.O. Box 226
Somers, NY, 10589

Action Math Assoc.
1358 Dalton Drive
Eugene, OR 97404

ACI Films, Inc.
35 West 45 St
New York, NY 10036

Acme Ruler Company Ltd
Foster Street
P.O. Box 239
Mt. Forest, Ontario,
Canada NOG 2LO

Acme United Corp.
100 Hicks Street
Bridgeport, CT 06609

Activity Resources Co.
Box 4875
Hayward, CA 94545

Addison-Wesley Pub. Co.
Reading, MA 01867

AERO Products
Dept. 194
St. Charles, IL 60174

Aims Instructional Media Services, Inc.
P.O. Box 1010
Hollywood, CA 90028

Allyn & Bacon, Inc.
470 Atlantic Ave.
Boston, MA 02210

American Association of School Librarians
50 East Huron Street
Chicago, IL 60611

American Book Co.
450 W. 33rd St.
New York, NY 10019

American Home Economics Association
2010 Massachusetts Avenue, NW
Washington, D.C. 20036

American Institute for Research
Metric Studios Center
P.O. Box 1113
Palo Alto, CA 94302

American National Metric Council
1625 Massachusetts Avenue, NW
Washington, D.C. 20036

American National Standards Institute
1430 Broadway
New York, NY 10018

Paul Amidon Assoc.
4329 Nicollet Ave. S.
Minneapolis, MN 55409

Ann Arbor Publishers Incorporated
P.O. Box 388
Worthington, OH 43085

Arco Publishing Co.
219 Park Avenue, South
New York, NY 10003

A.V. Instruction Systems
P.O. Box 191
Sommer, CT 06071

Baker & Taylor Company
6 Kirby Avenue
Somerville, NJ 08876

Barr Films
P.O. Box 7-C
Pasadena, CA 91104

Behavioral Research Laboratories
P.O. Box 577
Palo Alto, CA 94302

Bell & Howell
Audio Visual Products Division
7100 McCormik Road
Chicago, IL 60645

Beloit Tool Corp.
Rockton Road
South Beloit, IL 61080

Bennett Books·Co.
809 W. Detweiller Dr.
Peoria, IL 61614

BFA Educational Media
2211 Michigan Ave.
Santa Monica, CA 90404

BHU
23358 Hartland St.
Canoga Park, CA 91307

Brooks/Cole
Ralston Park
Belmont, CA 94002

R.W. Bruce Company
Educational Div. Dept. A
1401 Mt. Royal Avenue
Baltimore, MD 21217

Butterick Fashion Marketing Co.
161 Sixth Avenue
New York, NY 10013

California State Dept. of Education
P.O. Box 271
Sacramento, CA 95802

Canadian Metric Assoc.
P.O. Box 35
Fonthill, Ontario,
Canada LOS 1EO

Canadian Standards Association
178 Rexdale Boulevard
Rexdale, Ontario,
Canada M9W 1R3

The Center for Vocational Education
1960 Kenny Road
Columbus, OH 43210

Central Instrument Co.
Division of Impex Instrument Corp.
900 Riverside Drive
New York, NY 10032

Channing L. Bete Co. Inc.
45 Federal Street
Greenfield, MA 01301

Citation Press
50 West 44th St.
New York, NY 10036

Class Room Service Co.
P.O. Box 146
Campbell, CA 95008

Cliffs Notes, Inc.
Box 80728
Lincoln, NE 68501

Contemporary Ideas
P.O. Box 1703
Los Gatos, CA 95030

The Cooper Group
P.O. Box 728
Apex, NC 27502

Coronet Films
65 E. South Water St.
Chicago, IL 60601

Coronet Instructional Media
SSR Box 43
Weatherford, TX 76086

Creative Publications
P.O. Box 10328
Palo Alto, CA 94303

Creative Teaching Press
514 Hermon Vista Ave.
Montery Park, CA 91754

Creative Teaching Associates
P.O. Box 293
Fresno, CA 93708

The C-Thru Ruller Co.
6 Britton Drive
Bloomfield, CT 06002

Cuisenaire Company of America, Inc.
12 Church Street
New Rochelle, NY 10805

Damon/Educational Div.
80 Wilson Way
Westwood, MA 02090

Davidson Films
3701 Buchaner
San Francisco, CA 94123

Davis, A.R. and Co.
P.O. Box 24424
San Jose, CA 95154

Dick Blick Co.
P.O. Box 1267
Galesburg, IL 61401

Dominie Pty Ltd.
8 Cross Street
Brookvale, Australia 2100

Doubleday Multimedia
Box 11607
1371 Reynolds Ave.
Santa Ana, CA 92705

EdMediaTec, Inc.
P.O. Box 230
Wilkes-Barre, PA 18703

Edmund Scientific Co.
101 E. Glouchester Pike
Barrington, NJ 08007

Education Plus
18584 Carloyn Dr.
Costro Valley, CA 94546

Educational Activities Incorporated
P.O. Box 392
Freeport, NY 11520

Educational Aids Dept.
Union Carbide Research Center
P.O. Box 363
Tuxedo, NY 10987

Educational Aids and Supplies of Tomorrow
(E.A.S.T.), Inc.
P.O. Box 1337,
University Station
Gainesville, FL 32604

Educational Insights
20435 S. Tillman Ave.
Carson, CA 90746

Educational Metrics Corporation
207 Sunset Blvd.
Blacksburg, VA 24060

Educational Products, Incorporated
1211 W. 22nd Street
Oak Brook, IL 60521

Educational Science Consultants
P.O. Box 1674
San Leandro, CA 94577

Educational Teaching Aids
159 W. Kinzie St.
Chicago, IL 60610

Educational Tools, Inc.
901 W. Douglas
Wichita, KS 67213

Educulture, Inc.
1220 Fifth Street
Santa Monica, CA 90406

Encyclopedia Britannica Educational Corp.
425 North Michigan Ave.
Chicago, IL 60611

Enrich
3437 Alma Street
Palo Alto, CA 94306

ESP, Inc.
P.O. Drawer 5037
Jonesboro, AR 72401

ETA (Educational Teaching Aids)
159 West Kinzie St.
Chicago, IL 60610

Eye Gate House, Inc.
146-01 Archer Ave.
Jamica, NY 11435

Federal Reserve Bank of Minneapolis
Minneapolis, MN 55480

Films Incorporated
1144 Wilmette Avenue
Wilmette, IL, 60091

Filmstrip House
6633 W. Howard St.
Niles, IL 60648

Follett Library Book Company
1018 W. Washington Blvd
Chicago, IL 60607

Follett Publishing Co.
4300 West Ferdinand St.
Chicago, IL 60624

Franks Publishing & Printing Co.
P.O. Box 8026 University Station
Reno, NV 89507

Gamco Industries
P.O. Box 1911
Big Springs, TX 79720

Gel Sten Inc.
P.O. Box 2248
Palm Springs, CA 92262

General Learning Corp.
250 James Street
Morristown, NJ 07960

Mr. Anton Glaser
1237 Whitney Rd.
Southhampton, PA 18966

Goodyear Publishing Co.
15115 Sunset Blvd.
Pacific Palisades, CA 90272

Graphic Calculator Co.
234 James Street
Barrington, IL 60010

Great Ideas, Inc.
40 Oser Avenue
Hauppavege, NY 11787

Great Plains National Instructional
Television Library
Box 80669
Lincoln, NE 68501

Harcourt, Brace & Javanovich Inc.
757 Third Avenue
New York, NY 10022

Harper & Row Publishers, Inc.
10 East 53rd Street
New York, NY 10022

Hayes School Publishing
321 Penwood Avenue
Wilkinsburg, PA 15221

Holt, Rinehart & Winston, Inc.
383 Madison Ave.
New York, NY 10017

Houghton Mifflin Co.
One Beacon Street
Boston, MA 02107

Howard W. Sams & Company, Inc.
4300 West 62nd Street
Indianapolis, IN 46468

Idaho Research Foundation, Inc.
P.O. Box 3367
University Station
Moscow, ID 83843

Ideal School Supply Co.
11000 S. Lavergne Ave.
Oak Lawn, IL 60453

Imperial International Learning Co.
Box 548
Kankakee, IL 60901

Incentive Publications
Box 12522
Nashville, TN 37212

Industrial Press, Inc.
200 Madison Ave.
New York, NY 10016

Inquiry Audio Visuals
355 Lexington Avenue
New York, NY 10017

Instructor Curriculum Materials
Instructor Park
Dansville, NY 14437

Instructional Fair
P.O. Box 1650
Grand Rapids, MI 49501

International Tutors
22303 Devonshire St.
Chatsworth, CA 91311

Instructo/McGraw-Hill
Paoli, PA 19301

ITT Educational Publishing
Bobbs-Merrill Co.
4300 W. 62nd St.
Indianapolis, IN 46268

Interplanetary
P.O. Box 1338
E. Sausaleto, CA 94965

Janus Associates
P.O. Box 96
Wethersfield, CT 06109

Jay Scott Associates
P.O. Box 465
Memphis, TN 38104

Jaydee Specialties
P.O. Box 536
Wilmette, IL 60091

JEM Innovations
4568 E. 45th Street
Tulsa, OK 74135

J.J. Keller Assoc.
145 W. Wisconsin Ave.
Neenah, WI 54956

Kelm Manufacturing Co.
3151 U.S. 33 North
Benton Harbor, MI 49022

Kent Educational Serv.
Box 903
Oviedo, FL 32765

Knowledge Aid
6633 W. Howard St.
Niles, IL 60648

Laidlaw Brothers
Thatcher and Madison
River Forest, IL 60305

LaPine Scientific Co.
6001 South Knox Ave.
Chicago, IL 60629

Learning Arts
Dept. R.P.O. Box 917
Wichita, KS 67201

Learning Resource Center
10655 S.W. Greenburg Rd
Portland, OR 97223

Leicestershire Learning Systems
Box M74
New Gloucester, ME 04260

Library Filmstrip Center
3033 Aloma
Wichita, KS 67211

Listener
6777 Hollywood Blvd.
Hollywood, CA 90028

Little, Brown & Co. Inc.
34 Beacon Street
Boston, MA 02106

Love Publishing Co.
Dept. 12
6635 E. Villanova Pl.
Denver, CO 80222

MacLean-Hunter Learning Materials Co.
481 University Ave.
Toronto 101
Ontario, M5W
Canada

The Macmillan Co.
866 3rd Avenue
New York, NY 10022

MacMillan Publishing
Front and Brown Street
Riverside, NJ 08075

Martin Instrument Co.
19450 Grand River Ave.
Detroit, MI 84223

The Math Group
5626 Girard Ave, South
Minneapolis, MN 55419

Math-Master
P.O. Box 1911
Big Spring, TX 79720

McGraw-Hill
Webster Division
1221 Avenue of the Americas
New York, NY 10020

Media Materials Inc.
409 W. Cold Spring Lane
Baltimore, MD 21210

Meriwether Metrics
P.O. Box 1981
Montgomery, AL 36103

Chas. E. Merrill Publishing Co.
A Bell & Howell Co.
1300 Alum Creek Dr.
Columbus, OH 43216

Metric Association (U.S.)
Sugarloaf Star Route
Boulder, CO 80302

Metrication Institute of America
P.O. Box 236
Northfield, IL 60093

Metric Consultants
21720 W. North Avenue
Brookfield, WI 53005

Metrics Inc.
6140 Wayzata Blvd.
Golden Valley, NM 5416

Metric Supply International
1906 Main Street
Cedar Falls, IA 50613

Metric Teaching Aids
2858 Carolina Avenue
Redwood City, CA 94061

Metric Corporation
2500 Forsyth Rd
Orlando, FL 32807

Michigan Council of Teachers of Math
2165 E. Maple Rd
Birmingham, MI 48008

Milliken Publishing Co.
1100 Research Blvd
St. Louis, MO 63132

Milton Bradley Company
Educational Division
Dept. AT-N
Springfield, MA 01101

Minneapolis Public Schools
Math Basic Skills Development
2908 Colfax Ave., South
Minneapolis, MN 55408

Mississippi Authority for Educational
Television
P.O. Drawer 1101
Jackson, MI 39205

Modern Math Materials
1658 Albemarle Way
Burlingame, CA 94010

Moreland-Latchford Productions
299 Queen Street, W.
Toronto, Ontario

Moyer-Vico Limited
25 Milvan Drive
Weston, Ontario
MIL 121, Canada

Multi-Media Publishing, Inc.
1601 S. Federal Boul.
Denver, CO 80219

National Bureau of Standards
Metric Information Cntr
Washington, D.C. 20234

National Council of Teachers of Math
1906 Association Dr.
Reston, VA 22091

National Education Association
1201 16th Street, NW
Washington, D.C. 20036

National Microfilm Assoc.
Suite 1101
8728 Colesville Rd.
Silver Spring, MD 20910

National Science Teachers Association
1742 Connecticut Ave
Washington, D.C. 20009

National Textbook Co.
8259 Niles Center Rd.
Skokie, IL 60076

National Tool, Die & Precision Machining
P.O. Box 10344
Elmwood, CT 06110

National Tool Die Assoc.
9300 Liuraston Rd
Washington, D.C. 20022

Thomas Nelson, Inc.
30 East 42 Street
New York, NY 10017

Ohaus Scale Corporation
29 Hanover Rd.
Florham Park, NJ 07932

Orange County
Department of Education
1250 South Grand Avenue
Santa Ana, CA 92705

Ore Press
P.O. Box 1391
Cupertino, CA 95014

Osimiroid Company
2755 Woodshire Dr.
Hollywood, CA 90068

Parents' Magazine Press
52 Vanderbilt Avenue
New York, NY 10017

Parks and Math Company
945 Idlewood Dr.
San Jose, CA 95121

Pathescope Educational Films
71 Weyman Avenue
New Rochelle, NY 10802

Pauper Press
P.O. Box 303
Two Rivers, WI 54241

E. Joe Penn
4910 Carson Avenue
Indianapolis, IN 46227

J.C. Penny Co.
1301 Avenue of the Americas
New York, NY 10019

Perennial Educational
1825 Willow Road
P.O. Box 236
Northfield, IL 60093

Peterson Products
Route 4
Teegardon Rd.
Streator, IL 613364

Phi Delta Kappa
Eighth and Union
Box 789
Bloomington, IN 47401

Polymetric Services Inc.
18324 Oxnard St.
Tarzana, CA 93156

Prentice-Hall Learning Systems
P.O. Box 47X
Englewood Cliffs, NJ 07632

Prindle Webber & Schmidt
20 Newbury St.
Boston, MA 02116

Rand McNally Publishing Co.
Box 7600
Chicago, IL 60680

Random House
201 E. 50th Street
Rinehard Press
Corte Madra, CA 94925

Rapidesign, Inc.
Box 6039
Burbank, CA 91510

Real T Facts
P.O. Drawer 449
Warwick, NY 10990

Research Association
P.O. Box 13237
Gainsville, FL 32604

Robert C. Bellers & Associates, Inc.
131 Tulip Avenue
Floral Park, NY 11001

Roe International, Inc.
217 River Avenue
Patchogue, NY 11772

Rol-Ruler Company
1217 Dunham Road
Box 164
Rieglsville, PA 18077

Ronningen Metric Company
6102 Palo Cristi
Paradise Valley, AZ 8525

Rowsey Enterprises
P.O. Box 666
Friendswood, TX 77546

Roy G. Scarfo, Inc.
P.O. Box 217
Thorndale, PA 19372

Safco Manufacturing, Inc.
6500 Depot Drive
Box 7898
Waco, TX 76710

Sales Aids, Inc.
201 Bear Hill Road
P.O. Box 522
Waltham, MS 02154

Schloat Productions
A Prentice Hall Company
150 White Plains Rd.
Tarrytown, NY 10591

Science and Mathematics Teaching Center
College of Education
Publications Office
Erickson Hall
Michigan State University
East Lansing, MI 48824

Scholastic Magazines Incorporation
50 West 44th Street
New York, NY 10036

W. B. Saunders Co.
W. Washington Square
Philadelphia, PA 19105

Sadlier
300 Washington St.
Chicago, IL 60606

Sargent-Welch Scientific Co.
7300 N. Linde Ave.
Skokie, IL 60076

Sears Roebuck & Co.
Consumer Information Services
703 Sears Tower
Chicago, IL 60684

Selective Educational Equipment
3 Bridge Street
Newton, MA 02195

Science Research Associates
259 E. Erie Street
Chicago, IL 60611

Scott Foresman Co.
855 California Ave.
Palo Alto, CA 94304

Scott Resources, Inc.
Box 2121
Ft. Collins, CO 80521

Silver Burdett Co.
General Learning Corp.
4200 N. Industrial Blvd.
Indianapolis, IN 46254

The Smallstate Company
Box 796
Warwick, RI 02888

Society of Manufacturing Engineers
20501 Ford Road
Dearborn, MI 48128

See Hear Ltd.
44 Merer Street
Toronto, Canada M5E 1G9

Society for Visual Education
1345 Diversey Parkway
Chicago, IL 60614

South-Western Pub. Co.
5101 Madison Rd.
Cincinnati, OH 45227

Spectrum Educational Supplies Ltd.
8 Denison Street
Markham Ontario, Canada L3R 2P2

St. Regis
Consumer Products Division
3300 Pinson Valley Parkway
Birmingham, AL 35217

Stanley Tools
600 Myrtle Street
New Britain, CT 06050

Sterling Plastics
Borden Chemical
Sheffield St.
Mountainside, NJ 07092

Sterling Publishing Company, Inc.
419 Park Avenue S.
New York, NY 10016

Summit International Corporation
180 W. 2950 South
Salt Lake City, UT 34115

Swani Publishing Co.
Box 248
Roscoe, IL 61073

Teach 'Em Inc.
625 N. Michigan Ave.
Chicago, IL 60611

Teachers' Exchange of San Francisco
600 35th Avenue
San Francisc, CA 94121

Teacher's Supply Company
4281 Dacoma
Houston, TX 77018

Teaching Aids Co.
468 S. 300 West
Tooele, UT 84074

Teaching Resources Corp.
100 Boylston Street
Boston, MA 02116

Telex
9600 Aldrich Ave., S.
Minneapolis, MN 55420

Trend Enterprises, Inc.
P.O. Box 3073
St. Paul, MN 55165

Troubador Press
126 Folsom Street
San Francisco, CA 94105

Union Carbide
P.O. Box 363
Tuxedo Park, NY 10987

United States Air Force
Headquarters Air Training Command
Randolph Air Force Base, TX 78148

Universal Education and Visual Arts
100 University City Plaza
Universal City, CA 91608

Universal Supply Co.
1561 N. Bonnie Beach Place
Los Angeles, CA 90073

The Viking Press, Inc.
625 Madison Ave.
New York, NY 10022

Visual Instruction Productions
Victor Kayfetz
Productions, Inc.
295 W. 4th Street
New York, NY 10014

Vogel Book Company
P.O. Box 103
Bellevue, WA 98009

Jesse D. Wallace
1078 East 5th Avenue
Chico, CA 95926

Paul Wallach
2858 Carolyn Ave
Redwood City, CA 94061

Weber Costello Co.
1900 N. Narragansett Ave.
Chicago, IL 60639

West-East Bridge Unlimited
P.O. Box 1402
Scottsdale, AZ 85251

West Publishing Co.
50 W. Kellogg Blvd
St. Paul, MN 55102

Western Publishing Co.
850 Third Ave.
New York, NY 10022

Westminister Press
Witherspoon Building
Philadelphia, PA 19107

J. Weston Walch
321 Vally Street
Portland, ME 04104

Weybright and Tally
750 Third Avenue
New York, NY 10017

John Wiley & Sons, Inc.
605 Third Ave.
New York, NY 10016

Willowdale Media, Inc.
12 Parfield Drive
Willowdale, Ontario
Canada M2J 1B9

Willow House Publishers
P.O. Box 129
Stockton, CA 95201

Winston Press, Inc.
25 Groveland Terrace
Minneapolis, MN 55403

Worcester Public Schools
20 Irving Street
Worcester, MA 01609

WLW Enterprises
P.O. Box 43325
Lattjera Station
Los Angeles, CA 90043

Xerox College Publishing
275 Wyman Street
Waltham, MA 02154

Yoder Instruments
Negley, OH 44413

Zavell Math Co.
P.O. Box 36313
Los Angeles, CA 90036

Appendix F Conversion Tables

For nearly everybody, it is important that they do not try to convert metric measures to the customary measurement system and from the customary to the metric system. Trying to make these conversions is very discouraging and confusing for most people. However, it is important that they have a general idea of some approximate relationships. To help get a mental picture of metric measures, the following approximations may prove useful.

A kilometer is a little longer than a half mile.
A meter is about 10% longer than a yard.
A centimeter is a little shorter than a half inch.
A nickel weighs approximately 5 g.
A kilogram is about 10% heavier than 2 pounds.
A metric ton is about 10% heavier than a short ton.
A liter holds a little more than a quart.
A milliliter is about 1/5 of a teaspoon.
A hectare is about 2.5 acres.

REMEMBER THAT THE GOAL OF THE METRIC MOVEMENT IS TO GET PEOPLE TO *THINK METRIC*. For those people who may need to make conversions the following tables may prove useful.

LENGTH

Symbol	When You Know	Multiply By	To Find	Symbol
in	inches	2.5	centimeters	cm
ft	feet	30	centimeters	cm
yd	yards	0.9	meters	m
mi	miles	1.6	kilometers	km

AREA

Symbol	When You Know	Multiply By	To Find	Symbol
in²	square inches	6.5	square centimeters	cm²
ft²	square feet	0.09	square meters	m²
yd²	square yards	0.8	square meters	m²
mi²	square miles	2.6	square kilometers	km²
	acres	0.4	hectares	ha

MASS (WEIGHT)

Symbol	When You Know	Multiply By	To Find	Symbol
oz	ounces	28	grams	g
lb	pounds	0.45	kilograms	kg
	short tons (2000 lb)	0.9	metric tons	t

VOLUME

Symbol	When You Know	Multiply By	To Find	Symbol
tsp	teaspoons	5	milliliters	ml
tbsp	tablespoons	15	milliliters	ml
fl oz	fluid ounces	30	milliliters	ml
c	cups	0.24	liters	l
pt	pints	0.47	liters	l
qt	quarts	0.95	liters	l
gal	gallons	3.8	liters	l
ft³	cubic feet	0.03	cubic meters	m³
yd³	cubic yards	0.76	cubic meters	m³

TEMPERATURE (EXACT)

Symbol	When You Know	Multiply By	To Find	Symbol
°F	Fahrenheit temperature	5/9 (after subtracting 32)	Celsius temperature	°C

from metric measures

LENGTH

Symbol	When You Know	Multiply By	To Find	Symbol
mm	millimeters	0.04	inches	in
cm	centimeters	0.4	inches	in
m	meters	3.3	feet	ft
m	meters	1.1	yards	yd
km	kilometers	0.6	miles	mi

AREA

Symbol	When You Know	Multiply By	To Find	Symbol
cm^2	square centimeters	0.16	square inches	in^2
m^2	square meters	1.2	square yards	yd^2
km^2	square kilometers	0.4	square miles	mi^2
ha	hectares (10 000 m^2)	2.5	acres	

MASS (WEIGHT)

Symbol	When You Know	Multiply By	To Find	Symbol
g	grams	0.035	ounces	oz
kg	kilograms	2.2	pounds	lb
t	metric tons (1000 kg)	1.1	short tons	

VOLUME

Symbol	When You Know	Multiply By	To Find	Symbol
ml	milliliters	0.03	fluid ounces	fl oz
l	liters	2.1	pints	pt
l	liters	1.06	quarts	qt
l	liters	0.26	gallons	gal
m^3	cubic meters	35	cubic feet	ft^3
m^3	cubic meters	1.3	cubic yards	yd^3

TEMPERATURE (EXACT)

Symbol	When You Know	Multiply By	To Find	Symbol
°C	Celsius temperature	9/5 (then add 32)	Fahrenheit temperature	°F

more precise conversions

units of length

To Convert from Centimeters	
To	*Multiply by*
Inches	0.393 700 8
Feet	0.032 808 40
Yards	0.010 936 13
Meters	0.01

To Convert from Meters	
To	*Multiply by*
inches	39.370 08
feet	3.280 840
yards	1.093 613
miles	0.000 621 37
millimeters	1000
centimeters	100
kilometers	0.001

To Convert from Inches	
To	*Multiply by*
feet	0.083 333 33
yards	0.027 777 78
centimeters	2.54
meters	0.025 4

To Convert from Feet

To	Multiply by
inches	12
yards	0.333 333 3
miles	0.000 189 39
centimeters	30.48
meters	0.304 8
kilometers	0.000 304 8

To Convert from Yards

To	Multiply by
inches	36
feet	3
miles	0.000 568 18
centimeters	91.44
meters	0.914 4

To Convert from Miles

To	Multiply by
inches	63 360
feet	5280
yards	1760
centimeters	160 934.4
meters	1 609.344
kilometers	1.609 344

units of mass

To Convert from Grams

To	Multiply by
avoirdupois ounces	0.035 273 96
avoirdupois pounds	0.002 204 62
milligrams	1000
kilograms	0.001

To Convert from Kilograms

To	Multiply by
avoirdupois ounces	35.273 96
avoirdupois pounds	2.204 623
grams	1000
short tons	0.001 102 31
metric tons	0.001

To Convert from Metric Tons

To	Multiply by
avoirdupois pounds	2 204.623
short tons	1.102 311 3
kilograms	1000

To Convert from Avoirdupois Ounces

To	Multiply by
avoirdupois pounds	0.062 5
grams	28.349 523 125
kilograms	0.028 349 523 125

To Convert from Avoirdupois Pounds

To	Multiply by
avoirdupois ounces	16
grams	453.592 37
kilograms	0.453 592 37
short tons	0.000 5
metric tons	0.000 453 592 37

units of capacity, or volume,
liquid measure

To Convert from Milliliters

To	Multiply by
liquid ounces	0.033 814 02
liquid pints	0.002 113 4
liquid quarts	0.001 056 7
gallons	0.000 264 17
cubic inches	0.061 023 74
liters	0.001

To Convert from Liters

To	Multiply by
liquid ounces	33.814 02
liquid pints	2.113 376
liquid quarts	1.056 688
gallons	0.264 172 05
cubic inches	61.023 74
cubic feet	0.035 314 67
milliliters	1000
cubic meters	0.001
cubic yards	0.001 307 95

To Convert from Cubic Meters

To	Multiply by
gallons	264.172 05
cubic inches	61 023.74
cubic feet	35.314 67
liters	1000
cubic yards	1.307 950 6

To Convert from Liquid Pints

To	Multiply by
liquid ounces	16
liquid quarts	0.5
gallons	0.125
cubic inches	28.875
cubic feet	0.016 710 07
milliliters	473.176 473
liters	0.473 176 473

To Convert from Liquid Quarts

To	Multiply by
liquid ounces	32
liquid pints	2
gallons	0.25
cubic inches	57.75
cubic feet	0.033 420 14
milliliters	946.352 946
liters	0.946 352 946

To Convert from Gallons

To	Multiply by
liquid ounces	128
liquid pints	8
liquid quarts	4
cubic inches	231
cubic feet	0.133 680 6
milliliters	3 785.411 784
liters	3.785 411 784
cubic meters	0.003 785 411 784
cubic yards	0.004 951 13

units of capacity, or volume,
  dry measure

To Convert from Cubic Meters	
To	*Multiply by*
pecks	113.510 4
bushels	28.377 59

units of area

To Convert from Square Centimeters	
To	*Multiply by*
square inches	0.155 000 3
square feet	0.000 076 39
square yards	0.000 119 599
square meters	0.000 1

To Convert from Square Meters	
To	*Multiply by*
square inches	1 550.003
square feet	10.763 91
square yards	1.195 990
acres	0.000 247 105
square centimeters	10 000
hectares	0.000 1

To Convert from Hectares	
To	*Multiply by*
square feet	107 639.1
square yards	11 959.90
acres	2.471 054
square miles	0.003 861 02
square meters	10 000

To Convert from Square Inches	
To	*Multiply by*
square feet	0.006 944 44
square yards	0.000 771 605
square centimeters	6.451 6
square meters	0.000 645 16

To Convert from Square Miles

To	Multiply by
Square feet	27 878 400
Square yards	3 097 600
Acres	640
Square meters	2 589 988.110 336
Hectares	258.998 811 033 6

To Convert from Square Feet

To	Multiply by
square inches	144
square yards	0.111 111 1
acres	0.000 022 957
square centimeters	929.030 4
square meters	0.092 903 04

To Convert from Square Yards

To	Multiply by
square inches	1296
square feet	9
acres	0.000 206 611 6
square miles	0.000 000 322 830 6
square centimeters	8 361.273 6
square meters	0.836 127 36
hectares	0.000 083 612 736

To Convert from Acres

To	Multiply by
square feet	43 560
square yards	4840
square miles	0.001 562 5
square meters	4 046.856 422 4
hectares	.404 685 642 24

Reference

Units of Weight and Measure International (Metric) and U.S. Customary. NBS Miscellaneous Publication 286, U.S. Department of Commerce: May, 1967.

Index